Bacterial Protein Toxins

Bacterial toxins that act inside cells interact very specifically with key components of the cell, and some even manipulate the cell in subtle ways for their own purposes. These potent toxins, described in this book, will be of interest to both microbiologists and cell biologists. Some of these toxins are conventional multidomain toxins that are self-programmed to enter cells. Others are delivered by type III mechanisms, often as a package of potent molecules. The molecular targets for all these toxins mediate signal transduction and the cell cycle to regulate the crucial processes of cell growth, cell division, and differentiation. Thus, these potent toxins are not only responsible for disease but also provide a powerful set of tools with which to interrogate the biology of the cell. In addition, such toxins may act directly to promote carcinogenesis and, hence, their study is also of interest in a wider context.

ALISTAIR J LAX is Professor of Cellular Microbiology at King's College London, where he is Head of the Department of Microbiology within the Dental Institute. His research focuses on the novel mitogenic toxin of *Pasteurella multocida*, which activates several signalling pathways in the cell. He coauthored *Cellular Microbiology*, the first textbook on the subject, in 1999.

Over the past decade, the rapid development of an array of techniques in the fields of cellular and molecular biology has transformed whole areas of research across the biological sciences. Microbiology has perhaps been influenced most of all. Our understanding of microbial diversity and evolutionary biology, and of how pathogenic bacteria and viruses interact with their animal and plant hosts at the molecular level, for example, has been revolutionized. Perhaps the most exciting recent advance in microbiology has been the development of the interface discipline of cellular microbiology, a fusion of classic microbiology, microbial molecular biology, and eukaryotic cellular and molecular biology. Cellular microbiology is revealing how pathogenic bacteria interact with host cells in what is turning out to be a complex evolutionary battle of competing gene products. Molecular and cellular biology are no longer discrete subject areas but vital tools and an integrated part of current microbiological research. As part of this revolution in molecular biology, the genomes of a growing number of pathogenic and model bacteria have been fully sequenced, with immense implications for our future understanding of microorganisms at the molecular level.

Advances in Molecular and Cellular Microbiology is a series edited by researchers active in these exciting and rapidly expanding fields. Each volume will focus on a particular aspect of cellular or molecular microbiology, and will provide an overview of the area, as well as examining current research. This series will enable graduate students and researchers to keep up with the rapidly diversifying literature in current microbiological research.

Series Editors

Professor Brian Henderson
University College London

Professor Michael Wilson
University College London

Professor Sir Anthony Coates
St. George's Hospital Medical School, London

Professor Michael Curtis
St. Bartholemew's and Royal London Hospital, London

Published Titles

1. *Bacterial Adhesion to Host Tissues.* Edited by Michael Wilson 0521801079
2. *Bacterial Evasion of Host Immune Responses.* Edited by Brian Henderson and Petra Oyston 0521801737
3. *Dormancy and Low-Growth States in Microbial Disease.* Edited by Anthony R. M. Coates 0521809401
4. *Susceptibility to Infectious Diseases.* Edited by Richard Bellamy 0521815258
5. *Bacterial Invasion of Host Cells.* Edited by Richard J. Lamont 0521809541
6. *Mammalian Host Defense Peptides.* Edited by Deirdre Devine and Robert Hancock 0521822203

Forthcoming Titles in the Series

The Dynamic Bacterial Genome. Edited by Peter Mullany 0521821576
The Influence of Bacterial Communities on Host Biology. Edited by Margaret McFall-Ngai, Brian Henderson, and Edward Ruby 0521834651
The Yeast Cell Cycle. Edited by Jeremy Hyams 0521835569
Salmonella Infections. Edited by Pietro Mastroeni and Duncan Maskell 0521835046
Phagocytosis of Bacteria and Bacterial Pathogenicity. Edited by Joel Ernst and Olle Stendahl 0521845696
Quorum Sensing and Bacterial Cell-to-Cell Communication. Edited by Donald R. Demuth and Richard J. Lamont 0521846382

Advances in Molecular and Cellular Microbiology 7

Bacterial Protein Toxins
Role in the Interference with Cell Growth Regulation

EDITED BY
ALISTAIR J LAX
King's College London

CAMBRIDGE
UNIVERSITY PRESS

CAMBRIDGE UNIVERSITY PRESS
Cambridge, New York, Melbourne, Madrid, Cape Town,
Singapore, São Paulo, Delhi, Tokyo, Mexico City

Cambridge University Press
The Edinburgh Building, Cambridge CB2 8RU, UK

Published in the United States of America by Cambridge University Press, New York

www.cambridge.org
Information on this title: www.cambridge.org/9780521177467

First published 2005
First paperback edition 2011

A catalogue record for this publication is available from the British Library

Library of Congress Cataloguing in Publication data

Bacterial protein toxins : role in the interference with cell growth regulation / edited by
    Alistair J Lax.

    p. cm. – (Advances in molecular and cellular microbiology; 7)

    Includes bibliographical references and index.

    ISBN 0-521-82091-X (alk. paper)

    1. Bacterial toxins. I. Lax, Alistair, 1953– II. Series.

    QP632.B3B283 2004
    571.8′4 – dc22   2004055080

ISBN 978-0-521-82091-2 Hardback
ISBN 978-0-521-17746-7 Paperback

Additional resources for this publication at www.cambridge.org/9780521177467

Contents

List of Contributors   *page* ix
Preface   xiii

1 Toxins and the interaction between bacterium and host   1
  Alistair J Lax

2 The mitogenic *Pasteurella multocida* toxin and cellular
  signalling   7
  Gillian D Pullinger

3 Rho-activating toxins and growth regulation   33
  Gudula Schmidt and Klaus Aktories

4 Cytolethal distending toxins: A paradigm for bacterial
  cyclostatins   53
  Bernard Ducommun and Jean De Rycke

5 *Bartonella* signaling and endothelial cell proliferation   81
  Garret Ihler, Anita Verma, and Javier Arevalo

6 Type III–delivered toxins that target signalling pathways   117
  Luís J Mota and Guy R Cornelis

7 Bacterial toxins and bone remodelling   147
  *Neil W A McGowan, Dympna Harmey, Fraser P Coxon, Gudrun
  Stenbeck, Michael J Rogers, and Agamemnon E Grigoriadis*

8 *Helicobacter pylori* mechanisms for inducing epithelial
  cell proliferation   169
  Michael Naumann and Jean E Crabtree

9 Bacteria and cancer   199
   Christine P J Caygill and Michael J Hill

10 What is there still to learn about bacterial toxins?   227
    Alistair J Lax

Index   231

Plate section follows p. 146*
*These plates are available for download from
www.cambridge.org/9780521177467

Contributors

Klaus Aktories
Institut für Experimentelle und
Klinische Pharmakologie
und Toxikologie der
Albert-Ludwigs-Universität
Freiburg
Albert-Strasse 25
D-79104 Freiburg, Germany

Javier Arevalo
Proctor and Gamble America
Latina
PO Box 5578
Cincinnati, OH 45201, USA

Christine P J Caygill
UK National Barrett's Oesophagus
Registry
University Department of Surgery
Royal Free Hospital
Rowland Hill Street
London NW3 2PF, United Kingdom

Guy R Cornelis
Division of Molecular Microbiology,
Biozentrum
Universität Basel
Klingelbergstrasse 50–70
CH-4056 Basel, Switzerland

Fraser P Coxon
Department of Medicine and
Therapeutics
University of Aberdeen
Aberdeen AB25 2ZD
United Kingdom

Jean E Crabtree
Molecular Medicine Unit
St James's University Hospital
Leeds LS9 7TF, United Kingdom

Bernard Ducommun
LBCMCP-CNRS UMR5088
Université Paul Sabatier
Institut d'Exploration Fonctionnelle
des Génomes (IFR109)
118 route de Narbonne
31077 Toulouse, France

Agamemnon E Grigoriadis
Department of Craniofacial
Development
Dental Institute, King's College
London
Floor 27 Guy's Tower
Guy's Hospital
London SE1 9RT, United Kingdom

Dympna Harmey
The Burnham Institute
10901 North Torrey Pines Road
La Jolla, CA 92037, USA

Michael J Hill (deceased)
Formerly at European Cancer
    Prevention Organization
(UK) Headquarters
Nutrition Research Centre
South Bank University
103 Borough Road
London SE1 OAA, United Kingdom

Garret Ihler
Department of Medical
    Biochemistry and Genetics
Texas A&M College of Medicine
College Station, TX 77849, USA

Alistair J Lax
Department of Microbiology
Dental Institute, King's College
    London
Floor 28 Guy's Tower
Guy's Hospital
London SE1 9RT, United Kingdom

Neil W A McGowan
Department of Craniofacial
    Development
Dental Institute, King's College
    London
Floor 27 Guy's Tower
Guy's Hospital
London SE1 9RT, United Kingdom

Luís J Mota
Division of Molecular Microbiology,
    Biozentrum
Universität Basel

Klingelbergstrasse 50–70
CH-4056 Basel, Switzerland

Michael Naumann
Institute of Experimental Internal
    Medicine
Medical Faculty
Otto-von-Guericke-University
39120 Magdeburg, Germany

Gillian D Pullinger
Institute for Animal Health
Compton, Newbury, Berkshire
    RG 20 7NN, United Kingdom

Michael J Rogers
Department of Medicine and
    Therapeutics
University of Aberdeen
Aberdeen AB25 2ZD
United Kingdom

Jean De Rycke
UR 918 INRA de Pathologie
    Infectieuse et Immunologie,
BP 1, 37380 Nouzilly, France

Gudula Schmidt
Institut für Experimentelle und
    Klinische Pharmakologie
und Toxikologie der
    Albert-Ludwigs-Universität
    Freiburg
Albert-Strasse 25
D-79104 Freiburg, Germany

Gudrun Stenbeck
Bone and Mineral Centre
University College London
London WC1E 6JF
United Kingdom

CONTRIBUTORS

Anita Verma
Laboratory of Respiratory
  and Special Pathogens
DBPAP/CBER

Food and Drug Administration
Bethesda, MD 20892
USA

Preface

Many bacteria and higher eukaryotes live in harmony in a symbiotic relationship that benefits one or both of the partners. Indeed, we are colonised by bacterial cells which outnumber our own cells ten to one. This amicable bacterial lifestyle contrasts with a pathogenic one, in which the bacterium causes damage to its host. This is a potentially dangerous strategy for a bacterium, because the provoked host is capable of fighting back. A pathogenic lifestyle offers short-term gain. By outcompeting other bacteria within its host, the bacterium can achieve local dominance and, by more widespread colonisation, expand its territory more globally. However, evolution has to balance these advantages against the possibility that the bacterium is eliminated. The latter could occur if the bacterium is too weak to prevent its destruction by the strong host defences it has incited or if its potent virulence wipes out the host and, thus, its source of food.

As pathogenicity appears to be such a risky business, it may be an abnormal condition. Evidence to support this view comes from several lines. First, many of the genes involved in virulence appear to be relatively new – that is, new to the organism made pathogenic by their presence. These genes are frequently found on mobile genetic elements such as plasmids, phage, and pathogenicity islands that have recently been acquired by the organism. Secondly, it is thought that many severe diseases begin with ferocious virulence which later abates – mainly as a result of reduced pathogenicity – thereby suggesting that pathogenicity is not such a good strategy after all.

Several types of genes lead to pathogenicity. Bacterial adhesion, the topic of the first book in this series, is the first step a pathogen has to take. However, by itself adhesion does not promote disease: commensal bacteria adhere without causing tissue damage. The main route to bacterial disease is driven by bacterial toxins that specifically attack host cell function. Those toxins that act inside cells fall into two groups: the classical multidomain toxins and toxins that bacteria inject directly into cells. The latter are also called effector proteins. In each case, these toxins modify important cell functions. Because signalling mechanisms and the regulation of the cell cycle are key to a cell's continued existence, it is not surprising that these components have been targeted by toxins. Equally, it is not surprising that perturbation of these cellular systems frequently leads to aberrant regulation of cell growth.

This book describes toxins that interfere with the regulation of cell growth. Our perspective concerns not just the toxin, but also the cell and the whole organism. Several of the chapters describe the molecular interactions of toxins with their host target, whereas for other toxins these mechanisms are not yet clear. We have also included bacteria for which the precise effects on cell growth have not yet been ascribed to toxin action, on the premise that toxins are likely to be responsible. The relationship between the bacterial perturbation of cell growth and cancer is examined specifically for *Helicobacter pylori*, where various molecular mechanisms have been suggested and, generally, where the evidence is still at the epidemiological stage.

We hope that the reader will gain an understanding of these potent molecules and appreciate how their study not only illuminates infectious disease but also opens doors into exciting aspects of cell biology.

CHAPTER 1

Toxins and the interaction between bacterium and host

Alistair J Lax

The concept of a bacterial protein toxin was born in the 1880s as Friedrich Loeffler in Berlin, and Émile Roux and Alexandre Yersin in Paris, puzzled over the disease diphtheria. The bacteria were localised in the throats of patients and experimental animals, yet the disease caused systemic damage throughout the body. They reasoned that the bacteria must be producing a poison that could escape from the bacteria to cause widespread damage to the host. So the toxin concept was established right at the start of Medical Microbiology (Roux and Yersin, 1888), only a decade after Robert Koch had established the first definite link between a bacterium and disease with his seminal work on anthrax. However, it was only from the mid-twentieth century onwards that the action of any toxin was understood at the molecular level. Since then progress has been rapid, not only in our appreciation of the mode of action of historically known toxins but also in the discovery of new toxins with novel means of attacking cells.

CLASSES OF BACTERIAL PROTEIN TOXINS

The first toxin to be understood at the molecular level was one from *Clostridium perfringens*, a bacterium notorious for causing wound infections such as gas gangrene. This toxin is a phospholipase that attacks membranes of cells and, thus, it defined one of the three main categories of toxins, i.e., those that attack membranes (MacFarlane and Knight, 1941). The other group of toxins that attacks membranes contains the large number of toxins that insert into membranes to form pores – the pore-forming toxins. It is easy to envisage how these can damage the host cell, although it now transpires that the mode of action of these toxins is more complicated than was first thought.

Toxins of the second major category of toxins act on the cell surface and mimic the action of normal signalling molecules. These toxins are typified by the stable toxin (STa) from *Escherichia coli*. STa is a 19–amino acid peptide that mimics a natural hormone, guanylin, and binds tightly to its receptor in the intestine. This action chronically activates guanylate cyclase activity leading to raised cyclic GMP concentration in the cell and ultimately water exchange into the lumen of the gut and thus diarrhoea (Vaandrager et al., 1992).

Toxins of the third category of toxins enter the cell. The intracellularly acting toxins include such infamous toxins as cholera, diphtheria, and botulinum toxins. All of these attack key targets that are major players in the organisation of the cell and its ability to carry out its normal functions. Each is highly specific and interacts with only one target, or a highly related class of target. Toxins gain entry into the cell in one of two ways. The classical toxins, those that are released from the bacterium, are multidomain proteins that often comprise several subunits. Such toxins have to carry out three unrelated functions (Montecucco et al., 1994). First, they bind to the cell via a receptor to promote uptake into a membrane-bound vesicle. Second, the toxin, or part of it, has to cross the membrane into the cytosol. Finally, the toxin has to interact with the target. For all such toxins identified to date this interaction is an enzymatic one. This is one factor responsible for the extreme potency of these molecules. For example, one molecule of diphtheria toxin is sufficient to kill a cell (Yamaizumi et al., 1978). The other factor that makes these molecules so potent is target selection. Without exception such toxins modify and perturb proteins or processes in the cell that are crucial for its normal function.

Within the last decade or so, a new type of intracellularly acting toxin has been identified. These are generally called effector molecules, although it also seems entirely reasonable to refer to them as toxins. These are delivered directly into the cell by the bacterium, which makes a complex injection machine that forms a selective pore that crosses the two membranes of the bacterium and the membrane of the cell (Cornelis and Van Gijsegem, 2000). There are two slightly different delivery mechanisms that reflect the origin and evolution of these systems; they are referred to as either type III secretion systems (TTSS) or type IV secretion systems (TFSS). Much less is know about type IV secretion systems; in particular, very few of the effectors have been isolated so far (Ding et al., 2003).

Many bacteria inject toxic factors by TTSS, and it is generally the case that each bacterium delivers not one, but a cocktail of several effector proteins into the cell. Each of these toxin effectors has a different but key cellular target.

The discovery of these toxins solved the conundrum of how some of the deadliest bacteria, like *Yersinia, Shigella,* and *Salmonella,* were such potent pathogens, although they did not appear to produce classical toxins. Some of the effector toxins are enzymes like the classical intracellular acting toxins, but some mimic normal signalling molecules and bind to signalling molecules to affect their function in a transient manner.

TOXINS THAT MODULATE CELL FUNCTION

The traditional view of a toxin was that of a molecule that caused cellular damage and death at both the cellular and whole-animal level. While this is clearly true for some toxins, such as diphtheria toxin, a picture of toxin action is now emerging in which many toxins act on the signalling mechanisms in the cell in order to take over control of the cell rather than just kill it. This much more subtle tactic enables a pathogen to build a suitable environment for its survival and reproduction. With some bacteria, the preferred environment is intracellular, while other bacteria prefer to avoid phagocytosis.

The background knowledge necessary to understand the mode of action of many of the toxins described in this book has only recently become available, as cell biologists have unravelled the intricate signalling pathways that regulate the cell. Toxin science has greatly aided the advance of cell biology because of the great precision of intracellular acting toxins to pick out a limited set of molecular targets. All the targets chosen by these toxins are important proteins, the majority of which are involved in signalling, while some toxins appear to directly target the cell cycle. The high selectivity and precision of toxin action has in many cases led directly to the identification of signalling proteins and, furthermore, provides a set of valuable reagents for further analysis of these signalling molecules. Indeed, bacterial protein toxins are often referred to as the "cell biologist's toolkit." Toxins such as pertussis toxin and C3 toxin are routinely used by cell biologists to assess the involvement of their targets, the heterotrimeric G-protein G_i and the small G-protein Rho, respectively, in a particular process (Fiorentini et al., 1998; Albert and Robillard, 2002).

It has sometimes been difficult to reconcile the apparently sophisticated action of toxins with their role in disease. The general principle remains that any gene that conveys a competitive advantage to the bacterium will thrive, and it is clear that toxins that affect signalling pathways can aid the colonisation and establishment of the bacteria that express them in a number of ways. By killing cells, they can release a rich source of nutrients for the bacterium. In some cases, the toxin aids dissemination of the bacteria, such as with cholera. An increasing number of toxins are seen to target cells of the immune system, in particular those involved in innate immunity – the first line of defence against invading pathogens (Guldi-Rontani and Mock, 2002).

While the role of these toxins from the bacterial viewpoint is to aid colonisation of a human or animal host, the perturbation of host cell signalling can lead to various different outcomes at the cellular level that are dependent on both the cell type and signalling pathway affected. For example, toxins that affect proteins of the Rho family can affect the ability of the intoxicated cell to move (Oxford and Theodorescu, 2003). Some toxins can influence differentiation because that process is also controlled by signalling pathways. Similarly, signal transduction that is normally initiated by extracellular regulators that bind to cell surface receptors is intimately involved in the choice between apoptosis and cell growth and division, so it is not surprising that some of the toxins that interfere with signalling can affect that process as well. As a result, some toxins can induce or inhibit apoptosis, and at least one toxin is a potent mitogen.

The cellular signalling system has been honed by evolution over many years. The ability of the bacterium and its toxins to meddle with this finely tuned system that meticulously regulates cell function may carry dangers. Toxins that stimulate growth or apoptosis or inhibit differentiation in many ways behave like tumour promoters or inhibitors of tumour suppression (Lax and Thomas, 2002). This is of particular concern for chronic infections. The possible role of bacteria in carcinogenesis has a long and controversial history that began shortly after the linkage between bacteria and disease was established. There were numerous reports of the appearance of bacteria in tumours from the beginning of the twentieth century onwards. Many of these reports were not published in peer-reviewed journals and they were widely discounted by the mainstream scientific community at the time. Neither the longevity of the carcinogenic process nor the different stages of initiation, promotion, and progression were properly understood at that time, and this hampered a proper assessment of a possible bacterial role in cancer.

The discovery that *Helicobacter pylori* was involved not just in gastric ulceration but also in carcinogenesis forced a re-evaluation of the likely role of bacteria in this process (Parsonnet et al., 1991). It has recently become clear that other bacteria also affect cancer, most notably *Salmonella typhi* in people who become carriers and who, thus, are chronically infected with this bacterium. However, many controversies remain. In particular, the mechanism – or mechanisms – responsible are still being debated. Some scientists view prolonged inflammation caused by chronic infection as being the main contributing factor to carcinogenesis. However, others suggest more specific and direct effects may be implicated given the close similarities between the action of some toxins and tumour promoters, although there is currently no evidence that they have this role. Parallel work with viruses has shown that some viral infections predispose towards cancer, and the molecular mechanisms here are more clearly understood. Clearly, there is much more to be learned about the role of bacteria in cancer.

The chapters in this book cover not only toxins that explicitly disturb signalling pathways but also bacteria that impinge on cellular function in a similar manner. It is likely that these may well be found to express specific factors that explain these effects. In addition, the likely role of bacteria in the processes of carcinogenesis is discussed.

(5)

TOXINS, THE BACTERIUM AND ITS HOST

REFERENCES

Albert P R and Robillard L (2002). G protein specificity: Traffic direction required. *Cell. Signal.*, **14**, 407–418.

Cornelis G R and Van Gijsegem F (2000). Assembly and function of type III secretory systems. *Annu. Rev. Microbiol.*, **54**, 735–774.

Ding Z Y, Atmakuri K, and Christie P J (2003). The outs and ins of bacterial type IV secretion substrates. *Trends Microbiol.*, **11**, 527–535.

Fiorentini C, Gauthier N, Donelli G, and Boquet P (1998). Bacterial toxins and the Rho GTP-binding protein: What microbes teach us about cell regulation. *Cell Death Differ.*, **5**, 720–728.

Guldi-Rontani C and Mock M (2002). Macrophage interactions. *Curr. Top. Microbiol.*, **271**, 115–141.

Lax A J and Thomas W (2002). How bacteria could cause cancer – one step at a time. *Trends Microbiol.*, **10**, 293–299.

MacFarlane M G and Knight B C J G (1941). The biochemistry of bacterial toxins: The lecithinase activity of *Cl. welchii* toxins. *Biochem. J.*, **35**, 884–902.

Montecucco C, Papini E, and Schiavo G (1994). Bacterial protein toxins penetrate cells via a four-step mechanism. *FEBS Lett.*, **346**, 92–98.

Oxford G and Theodorescu D (2003). Ras superfamily monomeric G proteins in carcinoma cell motility. *Cancer Lett.*, **189**, 117–128.

Parsonnet J, Friedman G D, Vandersteen D P, Chang Y, Vogelman J H, Orentreich N, and Sibley R K (1991). *Helicobacter pylori* infection and the risk of gastric carcinoma. *New Engl. J Med.*, **325**, 1127–1131.

Roux Jr E and Yersin A (1888). Contribution a l'étude de la diphtherie. *Ann. Inst. Pasteur*, **2**, 620–629.

Vaandrager A B, Van der Wiel E, Hom M L, Luthjens L H, and de Jonge H R (1992). Heat-stable enterotoxin receptor/guanylyl cyclase C is an oligomer consisting of functionally distinct subunits, which are non-covalently linked in the intestine. *J. Biol. Chem.*, **269**, 16409–16415.

Yamaizumi M, Mekada E, Uchida T, and Okada Y (1978). One molecule of diphtheria-toxin fragment A introduced into a cell can kill cell. *Cell*, **15**, 245–250.

ALISTAIR J LAX

CHAPTER 2

The mitogenic *Pasteurella multocida* toxin and cellular signalling

Gillian D Pullinger

The *Pasteurella multocida* toxin (PMT) is produced by some type A and D strains of the Gram-negative bacterium *Pasteurella multocida*. These bacteria cause several animal infections and can occasionally cause human disease. PMT is the major virulence factor associated with porcine atrophic rhinitis, a non-fatal respiratory infection characterised by loss of the nasal turbinate bones and a twisting or shortening of the snout. However, PMT is highly toxic to animals, being lethal to mice at similar concentrations to diphtheria toxin. Despite these toxic properties, it turns out that PMT has unexpected effects on cells in culture leading to perturbation of several signalling pathways. The consequence of this action is that PMT can affect the regulation of cell growth and differentiation.

PMT IS A MITOGEN

The cellular effects of PMT have been most widely studied on Swiss 3T3 cells, a mouse fibroblast cell line. These cells are useful for studying growth factors since they are contact-inhibited and are readily quiesced by growing to confluence and allowing the cells to deplete the medium of growth factors. Rozengurt et al. (1990) first showed that PMT caused quiescent Swiss 3T3 cells to recommence DNA synthesis. The toxin is highly potent, inducing maximal DNA synthesis at only 1.25 ng/ml (or about 2 pM). This is equivalent to the DNA synthesis induced by 10% foetal bovine serum. Thus, PMT is more mitogenic for this cell type than any known growth factor (Figure 2.1, top panel). PMT also induces quiesced Swiss 3T3 cells to reinitiate proliferation (Figure 2.1, lower panel). The cells lose their density-dependent growth inhibition in the presence of 10 ng/ml toxin, and more than double in number by 4 days after addition of toxin to confluent monolayers. Addition

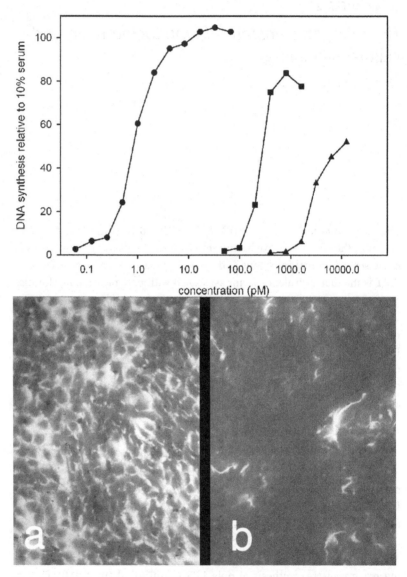

Figure 2.1. PMT is a mitogen for Swiss 3T3 cells. Top panel: relative mitogenicity of PMT
(●), platelet derived growth factor (PDGF) (■) and bombesin (▲); lower panel: cell
proliferation induced by 48 h PMT treatment: a, untreated cells; b, PMT treated cells.
(See www.cambridge.org/9780521177467 for color version.)

of PMT to subconfluent cells resulted in a striking proliferation of cells; the final saturation density increased 6-fold after treatment for 11 days. Thus the toxin induces more than one cycle of cell division. Furthermore, when quiescent cells were incubated for 24 h with 10 ng/ml PMT then trypsinised, washed, and replated in the absence of toxin, the subsequent growth of PMT-pretreated cells was markedly enhanced.

PMT is a mitogen for a number of other mesenchymal cells. Several murine cells lines including BALB/c 3T3, NIH 3T3, or 3T6 cells respond to PMT with a striking increase in cell growth. It is also mitogenic for tertiary cultures of mouse embryo cells and human fibroblasts (Rozengurt et al., 1990) and for primary embryonic chick osteoblasts (Mullan and Lax, 1996).

However, PMT does not appear to be a mitogen for certain cell types, such as embryonic bovine lung (EBL) or Vero cells. PMT affects the morphology of these cells, in particular causing cell rounding. This outcome has been regarded as a cytopathic or cytotoxic effect (Rutter and Luther, 1984; Pettit et al., 1993). However, it has been demonstrated that Vero cells treated with toxin remain viable by trypan blue exclusion assay, although they do not undergo DNA synthesis or proliferation (Wilson et al., 2000). The consequences of PMT treatment are therefore dependent on cell type.

PMT ACTS INTRACELLULARLY

There is considerable evidence that PMT is an intracellularly acting toxin. Typically, such toxins bind to a specific cell-surface receptor, and are taken up into cells by receptor-mediated endocytosis. This is followed by penetration through the membrane of the vesicle into the cell cytoplasm where the toxin interacts with a specific target protein. Many bacterial toxins studied to date target and enzymatically modify proteins with key roles in cellular functions such as cell-signalling and growth.

There is a lag period of at least an hour between application of PMT to cells and the onset of a cellular response (Rozengurt et al., 1990). This contrasts with receptor-acting growth factors whose mitogenic effects are seen within minutes. The longer lag is thought to be due to the requirement for toxin internalisation. Methylamine, an agent that increases endosomal and lysosomal pH and thereby inhibits receptor-mediated endocytosis, completely blocked the induction of DNA synthesis by PMT. Similarly, a neutralising antibody specific for PMT blocked the effect of the toxin when added shortly after toxin application, but was ineffective when added at 3 h. Finally, transient exposure of cells to PMT followed by extensive washing and incubation in toxin-free medium was sufficient to induce DNA synthesis. These experiments demonstrated that PMT acts intracellularly.

Binding and internalisation of PMT has been visualised by labelling purified toxin with colloidal gold (Pettit et al., 1993). Within 1 minute of addition of labelled PMT to Vero or osteosarcoma cells, gold-PMT particles were observed adhering to plasma membranes and in flask-shaped invaginations of the membrane. This binding was to a specific receptor, since it could be competed out by excess unlabelled toxin. After several minutes, the particles could be seen mainly in non-coated pits and in endocytic vesicles. The particles were not seen in vesicles deeper than 500 nm into the cytosol even after several hours. We have used PMT mutated in the catalytic domain in competition assays with wild-type toxin to demonstrate specific saturable binding with a Kd of 1.5 nM (Pullinger et al., 2001). The membrane receptor for PMT has not yet been identified, although binding of gold-labelled PMT was inhibited by mixed gangliosides (Pettit et al., 1993). Similarly, preincubation of PMT with gangliosides GM_1, GM_2 or GM_3 counteracted its effect on DNA synthesis (Dudet et al., 1996), suggesting that the receptor might be a ganglioside.

The translocation of toxins across membranes into the cytosol often involves a low pH processing step. For example, some toxins are proteolytically cleaved in endosomes before the catalytic fragment can be translocated into the cytosol (Olsnes et al., 1993). The finding that PMT action is blocked by the weak base methylamine suggests that PMT may enter the cytosol from an acidic compartment, such as an endosome or lysosome. PMT is highly resistant to proteolysis at neutral pH but becomes susceptible to a number of proteases at pH 5.5 and below, indicating that a conformational change occurs at this pH (Smyth et al., 1995). This was supported experimentally by transverse urea gradient gels and analysis of the circular dichroism profiles, each of which showed a transition in PMT structure at about this pH (Smyth et al., 1999). It is not yet known if PMT is cleaved by proteases *in vivo*.

PMT AFFECTS SEVERAL SIGNALLING PATHWAYS

PMT modifies cellular function by activating a number of cell signalling pathways. Intracellular toxins such as PMT act by subverting these signalling processes. Typically, they physically interact with and modify a specific signalling protein or group of related proteins and either activate or inactivate them. The effects are often long-lasting since the active components of toxins are enzymes. This contrasts with the normal transient activation of cell signalling pathways by receptor-acting growth factors. Thus, the cellular outcome of toxin action may be unusual.

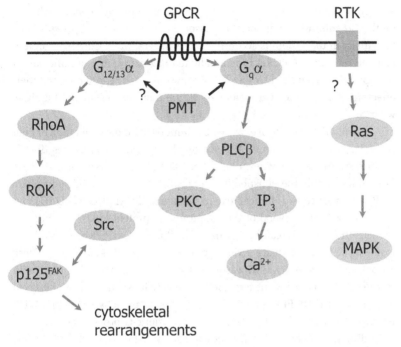

Figure 2.2. Summary of signalling pathways activated by PMT. See text for abbreviations.

PMT has been shown to activate three major signalling pathways (see Figure 2.2). It activates phospholipase C-β, which results in an increase in diacylglycerol, leading to activation of protein kinase C and increased production of inositol 1,4,5-trisphosphate (IP$_3$). The activation of this pathway is dependent on the heterotrimeric G-protein G$_q$. In addition, PMT activates the small GTPase, RhoA, an important protein involved in the regulation of the cytoskeleton and cell growth. Finally, PMT activates the mitogen-activated protein kinase pathway. These signalling pathways are each introduced in the following subsections, and the evidence for their modulation by PMT is discussed.

PMT Activates the G$_q$/Phospholipase C-β1 Pathway

The heterotrimeric guanine nucleotide binding proteins (G-proteins) are membrane-bound proteins that are composed of three subunits, α, β, and γ. They serve to transduce signals resulting from the binding of an agonist to its receptor on the cell surface to intracellular effector molecules (for reviews, see Gilman, 1987; Neer, 1995; Sprang, 1997). On the basis of sequence similarity the Gα subunits have been grouped into 4 classes (G$_s$, G$_i$, G$_q$, and G$_{12}$). The activity of G-proteins is regulated by cycles of guanine nucleotide exchange on the Gα subunit and dissociation/association into Gα and G$\beta\gamma$ subunits. Specificity in the system is accomplished in several ways. Certain G-protein coupled receptors (GPCRs) functionally interact with only a specific subset of G-proteins. There is also specificity between certain G-proteins and their effectors. For example, Gα_s interacts with and stimulates adenylyl cyclase, whereas PLC-β is activated by G$_{q/11}$.

Phospholipase C (PLC) enzymes are important for the generation of lipid secondary messengers implicated in signal transduction processes (for reviews, see James and Downes, 1997; Rhee, 2001). PLC catalyses the hydrolysis of a minor membrane phospholipid, phosphatidylinositol 4,5-bisphosphate (PIP$_2$) to produce two intracellular messengers, diacyl glycerol (DAG) and IP$_3$. DAG activates several of the subgroups of protein kinase C while IP$_3$ induces the release of calcium from intracellular stores. There are multiple isoforms of PLC, which are divided into four groups (β, γ, δ, and ε). These are activated differently. PLC-γ isoforms are direct targets of receptors with intrinsic, ligand-dependent tyrosine kinase activity such as platelet-derived growth factor (PDGF) receptors. PLC-β isoforms are activated by the GTP-bound α subunits of the G$_q$ class of heterotrimeric G-proteins as well as by G$\beta\gamma$ dimers. Receptors that are known to activate this Gα_q/PLC-β pathway include those for the mitogenic neuropeptides bombesin, endothelin, and vasopressin.

The first evidence that PMT activated PLC came from experiments in Swiss 3T3 cells, which showed that PMT treatment caused a dramatic increase in the production of total inositol phosphates (Rozengurt et al., 1990). This effect was time-dependent, occurring after a lag of 2–3 h and peaking at 5 h with a 25-fold increase in the level of inositol phosphates. The effect required uptake of the toxin into the cells, since it was blocked by methylamine. In contrast, PMT did not increase the intracellular level of the messenger cyclic AMP. Subsequent analysis of the inositol phosphates in PMT-treated cells by [^{3}H-inositol] labelling showed the increase was in the level of inositol 1,4,5-trisphosphate (IP$_3$) and its metabolic products (Staddon et al., 1991a). Furthermore, Ca^{2+} was mobilised from an intracellular pool. PMT treatment greatly reduced the ability of bombesin to mobilise Ca^{2+}, showing that PMT and bombesin mobilise Ca^{2+} from the same intracellular pool (Staddon et al., 1991a).

PMT at picomolar concentrations stimulated a 60% increase in the cellular content of diacylglycerol and of activated protein kinase C (Staddon et al., 1990), leading to phosphorylation of the 80 kDa MARCKS protein (myristoylated alanine rich C-Kinase substrate), a specific target for phosphorylation by protein kinase C. The toxin was shown to cause a translocation of protein kinase C to cellular membranes. These phosphorylation events were similar to those observed with bombesin. Cells depleted of protein kinase C by the addition of a high concentration of 4β-phorbol 12,13-dibutyrate failed to increase phosphorylation of the 80-kDa protein in response to PMT.

The data above demonstrated that PMT stimulates the phospholipase C-mediated hydrolysis of PIP_2. However, they do not distinguish between activation of the various PLC isoforms. To address this question, Murphy and Rozengurt (1992) pretreated cells with a low concentration of PMT to induce sub-maximal levels of IP_3 production, and then added either PDGF or one of the neuropeptides bombesin, vasopressin, or endothelin. PMT pretreatment caused a significant enhancement in the response to the neuropeptides but not to PDGF. This effect was greater than that expected for an additive effect. The rate of accumulation of IP_3 induced by bombesin was 2-fold higher in PMT-treated than untreated cells. Thus PMT selectively facilitated the PLC-β mediated signal transduction pathway utilised by neuropeptides. The enhancing effect of PMT was most likely due to a post-translational modulation of a component of the neuropeptide signal transduction pathway and not due to increased protein synthesis because cycloheximide had no effect. Indeed, we have recently shown that a non-mitogenic mutant of PMT also potentiates the action of bombesin, suggesting that potentiation is caused by binding of PMT to its target and not target activation (Baldwin et al., 2003). The addition of high concentrations of PMT to cells did not affect tyrosine phosphorylation of PLC-γ in either the absence or presence of PDGF. These results strongly suggest that PMT induction of IP_3 formation is via PLC-β. This view is supported by experiments using voltage-clamped *Xenopus* oocytes to measure PMT-mediated stimulation of PLC by monitoring the endogenous Ca^{2+}-dependent Cl^- current (Wilson et al., 1997). In this system injection of PMT induced a two-component Cl^- current similar to that produced by injection of IP_3. The PMT-induced current could be greatly reduced by prior injection with specific antibodies to PLC-β1. In contrast, antibodies to PLC-γ1 or PLC-β2 did not block the current. Antibodies to PLC-β3 caused a 25% decrease in the PMT response, suggesting the possibility that this phospholipase may play a minor role in the PMT response. However, these data suggested that PMT predominantly activates the β1 isoform of PLC.

The requirement for a functional G-protein for PMT activity was first demonstrated by experiments in which cells were permeabilised to allow the introduction of guanine nucleotide analogues into the cytosol. PMT retained its ability to induce increased IP_3 in cells permeabilised with streptolysin O. When such cells were treated with the G-protein antagonist GDPβS, the

PMT-induced accumulation of inositol phosphates was markedly inhibited (Murphy and Rozengurt, 1992). The non-hydrolysable agonist GTPβS reversed this inhibition. Thus PMT was shown to enhance the coupling of a G-protein to phospholipase C.

In accord with this, injection of antibodies to certain G-proteins inhibited the PMT-induced response in *Xenopus* oocytes (Wilson et al., 1997). Thus, antibodies against the common GTP-binding region of most G-protein α subunits (Gα_{pan}) and specifically against the C-terminal region of G$\alpha_{q/11}$ caused a pronounced reduction in the PMT-induced Cl$^-$ currents, whereas antibodies against α subunits of G$_s$ and G$_i$ classes of G-proteins had little effect. Furthermore, over-expression of Gα_q in *Xenopus* oocytes led to a 30- to 300-fold increased PMT response, whereas expression of Gα_q antisense cRNA reduced the response 7-fold. These data strongly suggested a role for G$_q$ in PMT activity. Interestingly, injection of the oocytes with antibody to the common region of Gβ subunits increased the PMT response by 4-fold. As mentioned earlier, PLC-β can be activated by G$\beta\gamma$ subunits most often when G$_i$ is activated. The most likely explanation for the increased response is that PMT acts through Gα_q and that this antibody sequesters $\beta\gamma$ dimers resulting in the presence of more free Gα_q, and that the free monomer is more susceptible to PMT than the heterotrimer.

Recently, the ability of PMT to function in Gα_q/Gα_{11} double-deficient fibroblasts, as well as in the corresponding single knockout cells, has been studied (Zywietz et al., 2001). The double knockout and the G$_q$ single knockout cells, were each unable to produce inositol phosphates in response to PMT. However, the G$_{11}$ single knockout did not prevent PMT-induced inositol phosphate formation. This shows that G$_q$ but not the closely related G$_{11}$ mediates the PMT-induced activation of PLC.

Activation of Rho and Rearrangement of the Actin Cytoskeleton

The Ras-related small G-proteins of the Rho family are implicated in regulation of cell growth and in organisation of the actin cytoskeleton (Chardin et al., 1989; Ridley and Hall, 1992). The Rho family contains several proteins including RhoA, Rac, and cdc42. These have different roles in cytoskeletal rearrangement: RhoA directs the formation of actin stress fibres and focal adhesions; Rac is required for membrane ruffling; and cdc42 is involved in filopodia formation. For a more detailed discussion, see Chapter 3. Small G-proteins resemble heterotrimeric G-proteins in that they are active when bound to GTP and inactivate when GDP-bound. In unstimulated cells, the majority of Rho is bound to guanine nucleotide dissociation factors (GDIs) and is in the cytosol (Sasaki and Takai, 1998). GTPase-activating proteins (GAPs) accelerate the intrinsic GTPase activity of Rho proteins, inducing their inactivation. In contrast, GEFs activate the small GTPases by catalysing the exchange of GDP for GTP. Many Rho-GEFs are regulated by heterotrimeric G-proteins. There is considerable evidence that $G\alpha_{12}$ and $G\alpha_{13}$ interact directly with the Rho-GEFs PDZ-RhoGEF and p115-RhoGEF (Hart et al., 1998; Kozasa et al., 1998; Fukuhara et al., 1999), and that these pathways are important for stress fibre formation. In addition, Rho is regulated by $G\alpha_q$, $G\alpha_i$, and $\beta\gamma$ dimers. These pathways are less well defined. It is possible that direct interaction with RhoGEFs occurs or regulation may be indirect via PKC, tyrosine kinases, or cyclic AMP. Activated Rho interacts with a number of downstream effectors. The best characterised of these is Rho kinase (p160/ROCK), which increases the extent of myosin light chain (MLC) phosphorylation (Kimura et al., 1996), contributing to actin cytoskeletal reorganization, cell adhesion, and migration. Responses for which the Rho effector is less well defined include DNA synthesis, cell growth, and gene transcription.

PMT treatment of Swiss 3T3 cells was shown to induce tyrosine phosphorylation of a number of key proteins (Lacerda et al., 1996). In particular, the non-receptor kinase p125[FAK] (focal adhesion kinase) and the cytoskeleton-associated adaptor protein paxillin were tyrosine phosphorylated. The phosphorylation of these two proteins was not affected by inhibitors of PKC or Ca^{2+} (GF109203X and thapsigargin, respectively). The phosphorylation of p125[FAK] in response to PMT occurred on Tyr[397] (Thomas et al., 2001), the major site for tyrosine autophosphorylation of p125[FAK], and potentially a high-affinity binding site for the SH2 domain of Src family proteins (Parsons and Parsons, 1997). Co-immunoprecipitation experiments showed that PMT treatment induced the formation of a stable FAK/Src complex, in which both kinases are believed to be active (Thomas et al., 2001).

Because p125[FAK] and paxillin both localise to focal contacts that form at the ends of actin stress fibres, the effect of PMT on the actin cytoskeleton was examined (Lacerda et al., 1996). Quiescent Swiss 3T3 fibroblasts contained disorganised actin, whereas the toxin induced the formation of thick bundles of parallel stress fibres (Figure 2.3). These formed after a lag of 1 h and reached a maximum at 8 h after PMT addition. The amount of vinculin increased in parallel, showing that PMT also induced focal adhesion assembly. Addition of cytochalasin D (a compound which disrupts the actin cytoskeleton) before PMT treatment blocked PMT-induced tyrosine phosphorylation of all phosphorylated proteins including p125[FAK]. This illustrates that the integrity of the actin cytoskeleton is necessary for p125[FAK] tyrosine phosphorylation to be activated by PMT. Microinjection of quiescent Swiss 3T3 cells

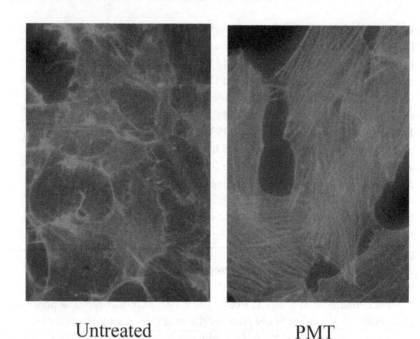

Untreated   PMT

Figure 2.3. Induction of stress fibres by PMT. Quiescent Swiss 3T3 cells were treated for
8 h with 20 ng/ml PMT, then the actin cytoskeleton was stained with fluorescently
labelled phalloidin. (See www.cambridge.org/9780521177467 for color version.)

with C3 exoenzyme, which ADP-ribosylates Rho to block its action, inhibited the tyrosine phosphorylation of p125[FAK]. Tyrosine phosphorylation of
p125[FAK] and the formation of stress fibres are dependent on the activity of
p160/ROCK, because two ROCK inhibitors (HA1077 and Y-27632) blocked
these events (Thomas et al., 2001). Thus, Rho and the Rho kinase are important for the activity of PMT on the cytoskeleton. Effects of PMT on cytoskeletal
reorganization and cell shape were also reported by Dudet et al. (1996). These
authors noted cell retraction as well as stress fibre formation in Swiss 3T3
cells. Both effects were blocked by methylamine, showing that they occurred
after toxin internalisation. Additionally, they described membrane ruffling of
PMT-treated cells, which was not blocked by methylamine. This was probably
a direct effect of the endocytosis event, and not a result of PMT activity.

The mechanism by which Rho is activated in response to PMT is unclear.
As described earlier, Rho can be activated by heterotrimeric G-proteins via
Rho-GEFs or other pathways. It was recently shown that PMT induces stress
fibre formation in $G\alpha_q/G\alpha_{11}$ double-deficient fibroblasts (Zywietz et al.,
2001). This stress fibre formation was dependent on Rho and Rho kinase

because microinjection with C3 or dominant-negative ROCK or treatment with a ROCK inhibitor blocked the actin rearrangements. Activation of Rho by PMT in wild-type cells and the double knockout cells was directly demonstrated by assaying its ability to bind a fusion protein consisting of GST fused to the Rho-binding domain of the Rho target rhotekin. These experiments showed that PMT-induced activation of Rho could occur in the absence of G_q or G_{11}. Furthermore, pre-incubation of cells with pertussis toxin did not affect PMT-mediated stress fibre formation, suggesting that G-proteins of the G_i class were not required. Other possible mechanisms for Rho activation such as activation by $G_{12/13}$ or by tyrosine kinases remain to be investigated.

Studies by Essler et al. (1998) using confluent endothelial cells (HUVECs) showed that actin reorganisation and cell retraction caused by PMT led to a 10-fold increase in transendothelial permeability. This was almost completely abolished by pretreatment with exoenzyme C3, showing that Rho was involved. The effect of C3 was reversed by addition of the myosin light chain phosphatase (MLCP) inhibitor tautomycin. It is thought that the changes in permeability are linked to PMT induced cytoskeletal changes in these cells. The prominent stress fibres were abolished if cells were microinjected with either the RBD (Rho binding domain) or PH (Pleckstrin homology) domain of Rho kinase prior to PMT addition. Microinjection of cells with constitutively active MLCP completely abolished PMT-induced stress fibre formation. These results are consistent with the idea that PMT decreases MLCP activity via Rho/Rho kinase and that this results in increased phosphorylation of MLC. Protein phosphatase assays confirmed that PMT inactivates MLCP. It was further demonstrated that MLC kinase was required for actin reorganisation, because the MLC kinase inhibitor KT5926 blocked actin rearrangements, and that MLC was phosphorylated in a time-dependent manner by PMT. The endothelial cell contraction induced by this pathway was suggested to contribute to the vascular permeability, oedema, and emigration of neutrophils that are observed during infection with toxigenic *P. multocida*.

The growth of most normal cells requires contact with an adhesive substratum to proliferate, a requirement that is removed by oncogenic transformation (Varmus, 1984). The ability of PMT to stimulate anchorage-independent growth was assessed in Rat-1 fibroblasts, a cell line that can readily be induced to form colonies in semisolid medium. PMT at picomolar concentrations was found to potently induce an increase in the formation of colonies (Higgins et al., 1992). The magnitude of the effect was greater than that achieved by nanomolar concentrations of epidermal growth factor or platelet-derived growth factor. Integrin-mediated signals, including tyrosine phosphorylation of focal adhesion proteins, have been implicated in promoting anchorage-independence. It is likely that PMT circumvents the requirement for integrin-mediated signals generated in adherent cells as a result of its considerable and persistent induction of stress fibres and tyrosine phosphorylation of focal contact proteins.

Activation of Mitogen-Activated Protein Kinase Cascades by PMT

The mechanisms whereby PMT induces its mitogenic effects are not fully understood, although PMT is known to activate the Erk mitogen-activated protein kinase (MAP kinase) pathway. Many heterotrimeric G-proteins are known to activate this pathway. This can be achieved by a mechanism involving activation of receptor or non-receptor tyrosine protein kinases and Ras/Raf, or by a poorly defined pathway involving stimulation of PKC isoforms (Gutkind, 1998).

Ras is a small membrane-associated G-protein, which is active in its GTP-bound state. Activated Ras mediates some of the intracellular signals normally seen in response to mitogens that bind to receptor tyrosine kinases. It can also be activated via the non-receptor tyrosine kinase p125[FAK]. GTP-binding to Ras is regulated by specific guanine nucleotide exchange factors including Sos (Son of Sevenless). The best characterised effector of Ras is Raf, a serine/threonine kinase that links Ras to the Erk MAP kinase cascade (Raf/Mek/Erk). This cascade spans from the plasma membrane to the nucleus. Erk phosphorylates and regulates key enzymes and also translocates to the nucleus, where it plays a role in regulating expression of genes essential for proliferation, for example, c-*myc*. In addition to this pathway, heterotrimeric G-proteins also induce a distinct pattern of expression of immediate early genes, including those of the *jun* and *fos* family. The activity of *jun* is regulated by a novel family of MAP kinases named Jun kinases (Jnks). Jnks are linked to GPCRs via the small GTPases Rac and cdc42.

PMT was shown to stimulate Erk1/2 phosphorylation in Swiss 3T3 cells after a 4-h incubation (Lacerda et al., 1997). It also stimulated phosphorylation of Erk1/2 in HEK 293 cells to reach a maximum of 6- to 10-fold above basal level after 12–48 h of continuous exposure (Seo et al., 2000). To determine whether Erk1/2 phosphorylation occurred downstream of G_q activation, the effect of PMT pre-treatment on Erk1/2 phosphorylation induced by the $G_{q/11}$-coupled α-thrombin receptor was assayed. PMT pre-treatment failed to produce an additive response, suggesting that a common G-protein pool mediated both effects. In contrast, PMT pre-treatment produced an additive Erk1/2 phosphorylation with the G_i-coupled LPA receptor, indicating that G_i was not involved in the PMT response. Interestingly, the PMT effect was not additive with the EGF receptor. These results were confirmed using specific inhibitors of G-protein function. Erk activation by PMT was unaffected by the G_i inhibitor, pertussis toxin, but was significantly attenuated by the expression of $G_{q/11}$ inhibitory peptides (the C-terminal fragment of $G\alpha_q$, $G\alpha_q$-(305–359), and an inactive mutant of the G-protein-coupled receptor kinase GRK2, GRK2(K2220R)). These results demonstrated that PMT induced Erk1/2 activation via $G_{q/11}$ activation.

The pathway linking G_q activation by PMT to Erk1/2 phosphorylation was also investigated (Seo et al., 2000). The role of PKC was assessed by the use of PKC inhibitors. The inhibitors GF109203X and Ro31-8220, which inhibit classical isoforms of PKC, had no effect on PMT-induced Erk phosphorylation. In contrast, the tyrphostin AG1478, which inhibits signalling via the EGF receptor, profoundly inhibited ERK phosphorylation in response to PMT. These data indicate that PMT stimulation of the Erk pathway is not mediated by PKC, but involves ligand-independent transactivation of the receptor tyrosine kinase, the EGF receptor. The expression of dominant inhibitory mutants of mSos and Ha-Ras significantly inhibited PMT-mediated Erk activation, providing further evidence for this.

The activation of this pathway by PMT is consistent with earlier findings that exposure of Swiss 3T3 cells to PMT resulted in the loss of cell surface EGF receptors – an effect that may represent EGF receptor downregulation following PMT-induced transactivation (Staddon et al., 1990). This was demonstrated by showing that PMT-treated cells bound less ^{125}I-EGF. The decreased binding occurred after a lag of at least 1 h and was sensitive to methylamine, indicating that PMT internalisation was an essential prerequisite. When PKC was down-regulated by pre-treatment with 4β-phorbol 12,13-dibutyrate, transactivation of the EGF receptor in response to bombesin was completely blocked but that induced by PMT was inhibited by only about 50%. These results indicate that PMT transmodulated the EGF receptor by both PKC-dependent and -independent pathways. The discrepancy between this suggested role for PKC and the lack of a role for PKC in Erk activation indicated by the results of Seo et al. (2000) may reflect the different cell types being used. Thus, the pathways leading to cell proliferation may differ between cells.

Recent experiments have shown that PMT activates the MAP kinase, Jnk, in fibroblasts (Zywietz et al., 2001). These authors also showed that activation of Erk and Jnk in response to PMT occurred in fibroblasts deficient in $G\alpha_q$ and $G\alpha_{11}$. This contrasts with the experiments using G_q inhibitors (Seo et al., 2000) that implicated G_q in Erk activation, and clearly showed that there is an alternate pathway not involving G_q or G_{11}. It is possible that Erk activation occurs via G_q and another mechanism.

EFFECT OF CELL TYPE ON OUTCOME OF PMT-MEDIATED ACTIVATION OF SIGNALLING PATHWAYS

As discussed earlier, PMT is mitogenic for some cell types but not for others. This is probably due to differential expression or regulation of signalling molecules in different cells. As an example, one study compared the effects of PMT in Swiss 3T3 cells with those in Vero cells (Wilson et al., 2000). PMT induced the up-regulation of cell cycle markers in Swiss 3T3 cells. Expression of the cyclins D1, D2, D3, and E, p21, PCNA, c-myc, and Rb/107 were all increased by addition of PMT to quiescent Swiss 3T3 fibroblasts. These effects are consistent with PMT induction of cell cycle reentry from G_0 into G_1 followed by progression through G_1, S, G_2, and M. Under the experimental conditions used, the cell cycle arrested after two or three rounds in mid to late G_1. In contrast in Vero cells, which were not induced to proliferate by PMT, several of the cell cycle markers (PCNA and cyclins D3 and E) were not up-regulated.

In one recent study, PMT has been used to elucidate the signalling pathways that promote cardiomyocyte hypertrophy (Sabri et al., 2002). Myocardial hypertrophy occurs as a result of stresses that increase cardiac work, but in the long term generally progresses to cardiac failure. The G_q family plays a role in hypertrophy. Thus, modest increases in wild-type $G\alpha_q$ have previously been shown to induce stable cardiac hypertrophy, whereas very intense $G\alpha_q$ stimulation induced dilated cardiomyopathy, with functional decompensation and cardiomyocyte apoptosis (Adams et al., 1998). This study used PMT as a tool to mimic these effects in cardiomyocytes and study the signalling pathways involved. It was found that PMT stimulated nPKC isoforms (PKCδ and PKCε), and this led to activation of MAP kinase cascades, which is consistent with induction of hypertrophy. In contrast to HEK 293 cells (Seo et al., 2000) and to cardiac fibroblasts (Sabri et al., 2002), activation of MAP kinase was independent of EGF receptor activation. Activation of nPKC isoforms in cardiomyocytes also resulted in decreased Akt phosphorylation. Akt is a serine threonine kinase, which when activated by phosphorylation acts as a cell survival factor. Thus the repression of Akt activation by PMT is consistent with the increased apoptosis observed in cardiomyocytes.

Once the molecular target(s) of PMT are established, the toxin will be a valuable research tool for analysing signalling pathways in different tissues and cells.

THE TARGET OF PMT AND ITS MODIFICATION

It is clear that PMT acts via the heterotrimeric G-protein, G_q, and this protein is a possible primary target of the toxin. The signalling pathways stimulated by PMT could theoretically all be activated via G_q. However, the use of $G\alpha_q/G\alpha_{11}$ deficient fibroblasts clearly demonstrates that PMT can still exert some of its effects in the absence of G_q. Therefore, G_q cannot be the sole target. Toxins in general are highly specific in their mode of action, modifying a single target or small group of closely related proteins. For example, cholera toxin modifies both G_s and G_i. Therefore, it seems likely that a second PMT target could be closely related to G_q, for example another heterotrimeric G-protein such as G_{12} or G_{13}. PMT induces actin stress fibre formation, Rho activation, and phosphorylation of MAP kinases in the $G\alpha_q/G\alpha_{11}$ knockout cells – effects that could be induced by activation of a member of the G_{12} class of G-proteins. However, there has not yet been any experimental demonstration that PMT modifies G_q or another heterotrimeric G-protein. It remains a possibility that the target could be something else. For example, it has been suggested that a guanine nucleotide exchange factor could be modified by PMT (Zywietz et al., 2001). Whatever the primary target(s) are, PMT is the first toxin identified that activates G_q, and is therefore a useful reagent for investigating G_q-mediated signalling pathways.

Recent results from our laboratory (Baldwin et al., 2003) have shown that $G\alpha_q$ is tyrosine phosphorylated in PMT-treated cells. This phosphorylation was time- and dose-dependent, and required internalisation of the toxin. Surprisingly, an inactive point mutant of PMT (C1165S) also phosphorylated $G\alpha_q$ with similar kinetics. These results are consistent with a mechanism in which PMT interacts physically with $G\alpha_q$. Thus, the catalytically inactive mutant toxin is proposed to bind to $G\alpha_q$ and induce its phosphorylation perhaps by altering its conformation, but is unable to enzymatically modify and activate it.

The enzymatic activity of PMT remains to be identified. The catalytic domain of PMT (see below) is not significantly homologous to known proteins that might give an indication of its function. Other toxins for which a mode of action has been determined have a limited number of enzymatic activities, including ADP-ribosylation, deamidation, transglutamination, glycosylation, and proteolysis. It is possible that PMT may possess a novel enzymatic activity. PMT is not an ADP-ribosyltransferase because labelling of cells with [$2-^3$H]adenine prior to PMT treatment did not result in increased labelling of any protein bands (Staddon et al., 1991b). Furthermore, mutation of a possible ADP-ribosylation motif in PMT had no effect on the activity of the toxin

(Ward et al., 1994). The investigation of other possible activities of PMT has not been reported.

THE STRUCTURE OF PMT

The gene encoding PMT was cloned from toxigenic *P. multocida* in three different laboratories (Petersen and Foged, 1989; Kamps et al., 1990; Lax and Chanter, 1990), and its complete nucleotide sequence determined (Buys et al., 1990; Lax et al., 1990; Petersen, 1990). The toxin gene was recently found to be located on a lysogenic prophage, indicating that it was horizontally acquired (Pullinger et al., 2004). The gene encodes a protein of 1285 amino acids (approximately 146 kDa). A protein of this molecular weight has been purified to homogeneity from recombinants containing the whole PMT gene, and shown to possess biological activity.

Large intracellular toxins are generally composed of functional domains with separate roles in receptor-binding/membrane translocation (the binding or "B" component) and in enzymatic modification of the target (the active "A" component). Analysis of the PMT sequence showed significant homology of the N-terminal of PMT with the N-terminal of the cytotoxic necrotizing factors (CNF1 and 2) of *E. coli*. In particular, the region of highest homology corresponds to residues 230 to 530 of the PMT sequence (Pullinger et al., 2001). This homologous region includes a predicted hydrophobic, helical region (residues 379 to 498) that is a potential membrane translocation domain. In CNF the N-terminus is known to function in binding and internalisation of the toxin (Lemichez et al., 1997), suggesting a similar role for the N-terminus of PMT. Recent analysis of this region in PMT supports this prediction (Baldwin et al., 2004). Secondary structure prediction analysis suggested that the C-terminal of PMT (residues 889 to 1220) is an alternating α/β fold, which is likely to fold into a structurally discrete domain. This prediction was supported by marginal homology of this region to proteins of known mixed α/β structure (Pullinger et al., 2001).

The first study to analyse which parts of the toxin molecule were required for activity used a deletion strategy (Petersen et al., 1991). Deletions were made using convenient restriction sites. Four toxin deletants were stable and expressed in high enough amounts for purification and functional analysis. Three of these were completely inactive, and one (lacking residues 505–568) retained some activity, suggesting this region was not critical for activity. This region is immediately downstream of the putative hydrophobic helical region. Site-directed mutagenesis has subsequently been employed to identify important residues. Many large toxins use disulphide bonds to link or stabilize multiple-domain structures and to enable the delivery of the catalytic fragment to the cytosol, so cysteine residues are often significant. All 8 cysteines of PMT were individually mutated and the activity of the variant toxins assessed (Ward et al., 1998). Only the most C-terminal of the 8 cysteines (Cys 1165) was essential for activity. Mutant C1165S was not cytotoxic for EBL cells, lacked mitogenic activity for Swiss 3T3 cells and produced no discernible effect in gnotobiotic piglets even when given at $1000\times$ the wild-type LD_{50}. The loss of activity of C1165S was not due to gross structural changes because it displayed similar protease resistance and circular dichroism spectra to wild-type toxin (Ward et al., 1998). Mutant C1165S retained its ability to bind to cells because it could block the activity of wild-type PMT (Pullinger et al., 2001). Thus, C1165 was proposed to be essential for the enzymatic activity of PMT. We have recently evaluated the role of the four most C-terminal histidine residues using a similar approach, and found that H1205 is also essential for enzymatic activity (unpublished results). Mutation of H1223 also significantly decreased toxin activity. Similarly, Orth et al. (2003) found that H1205 and H1223 are essential for activity.

The functional domains of PMT have recently been located by analysis of purified PMT peptides. Results from our laboratory and from an independent study both localised the catalytic domain to the C-terminal. Thus, we showed that microinjection of a C-terminal peptide consisting of amino acids 681–1285 into quiescent Swiss 3T3 fibroblasts induced DNA synthesis and led to morphological changes typical of PMT (Figure 2.4; Pullinger et al., 2001). Microinjection of N-terminal peptides had no effect. Furthermore, microinjection of antibodies against this C-terminal fragment inhibited the activity of wild-type toxin added subsequently to the medium. Similarly, Busch et al. (2001) used electroporation to introduce peptides into EBL cells, and found that the slightly larger C-terminal peptide (581–1285) caused reorganisation of the actin cytoskeleton and rounding of the cells, resembling the effect of PMT on these cells. N-terminal peptides were ineffective. Electroporation of the 581–1285 peptide also led to accumulation of inositol phosphates in EBL cells. A smaller C-terminal peptide (residues 701–1285) was inactive in these assays.

An N-terminal PMT peptide (residues 1–506) competed with full-length toxin for binding to cell surface receptors, and therefore contains the binding domain (Figure 2.4; Pullinger et al., 2001). This peptide probably also contains the membrane translocation domain. Consistent with this concept, antibodies raised against this peptide bound efficiently to native PMT, suggesting that the N-terminus is surface-located.

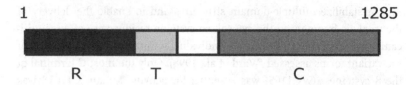

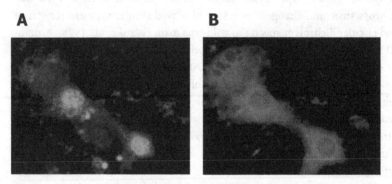

Figure 2.4. The functional domains of PMT. Top panel: diagram showing the approximate locations of the functional domains. R, receptor-binding domain; T, membrane translocation domain; C, catalytic domain. Lower panel: quiescent Swiss 3T3 cells were microinjected with the C-terminal of PMT (residues 681–1285) and with rabbit IgG. DNA synthesis was assayed by addition of BrdU, which is incorporated into the DNA of activated cells. A, green nuclei represent BrdU positive cells; B, microinjected cells stained red. (See www.cambridge.org/9780521177467 for color version.)

In contrast, experiments by Wilson et al. (1999) indicated that the N-terminus had biological activity. This interpretation was based on the microinjection of peptides into *Xenopus* oocytes. An N-terminal peptide consisting of residues 1 to 568 induced a Ca^{2+}-dependent Cl^- current. However, because this peptide includes the hydrophobic helical region thought to be the translocation domain, the result might have been an artefact of membrane insertion.

Further mutagenesis analysis will be needed to locate the minimal functional domains and to identify residues essential for specific functions.

PMT AND DISEASE

Atrophy of the nasal turbinate bones and twisting or shortening of the snout characterise atrophic rhinitis (AR) of pigs (Switzer and Farrington, 1975).

The severe disease is caused by respiratory infections with both *P. multocida* and *Bordetella bronchiseptica* (Pedersen and Barfod, 1981). In gnotobiotic pigs, infection with *B. bronchiseptica* alone caused moderate atrophy but the bones regenerated (Rutter et al., 1982), whereas in mixed infections with toxigenic *P. multocida* colonisation by large numbers of the pasteurellae occurred and severe, progressive disease was produced (Rutter and Rojas, 1982). Gnotobiotic pigs infected with *P. multocida* alone were colonised by far fewer bacteria, showing that prior colonisation with *B. bronchiseptica* led to conditions suitable for *P. multocida* colonisation (Rutter and Rojas, 1982). Inoculation of bacteria-free extracts from toxigenic *P. multocida* strains or of purified PMT into pigs produced turbinate atrophy (Rutter and Mackenzie, 1984; Chanter et al., 1986). Thus PMT is responsible for the turbinate atrophy caused by *P. multocida* in AR. The toxin has also been shown to be lethal for mice after intra-peritoneal inoculation and produces dermonecrotic skin lesions in mice or guinea pigs injected intra-dermally (Nakai et al., 1984).

PMT is mitogenic for osteoblasts, the cells that lay down bone (Mullan and Lax, 1996). It also down-regulates the expression of several markers of osteoblast differentiation, and so is thought to inhibit the differentiation of immature osteoblasts into mature osteoblasts. The toxin might also act on osteoclasts. These results and their relevance to possible mechanisms for the bone loss caused by PMT are discussed in Chapter 7.

As well as the effects on bone, intraperitoneal introduction of PMT led to proliferative changes in the epithelium of the bladder wall and ureter (Rutter and Mackenzie, 1984; Lax and Chanter, 1990). A similar effect was also observed following nasal infection of gnotobiotic pigs with a toxigenic *P. multocida* strain (Figure 2.5; Hoskins et al., 1997). Hyperplasia and vacuolation of the transitional epithelial lining in the bladder occurred, without any apparent evidence of inflammation. The bladder epithelium was 4- to 6-fold thicker than in a control animal, and this was due to increased cell number rather than cell enlargement. In another *in vivo* experiment, rat osteosarcoma cells were implanted subcutaneously into nude mice, and the effect of a course of PMT injections was assessed (Dyer et al., 1998). After implantation all animals developed tumours. The tumours in mice given three injections of PMT were consistently larger than those in control mice receiving no toxin. The results indicated that PMT had a mitogenic effect and contributed to neoplastic growth. In addition, the enhanced tumour growth led to an increased incidence of metastasis.

These results raise concern that carriage of toxigenic *P. multocida* could lead to systemic proliferative changes. This could occur in adult pigs where carriage of toxigenic *P. multocida* does not cause AR. *P. multocida* also causes human infections. Strains isolated from wounds inflicted by dogs or cats are

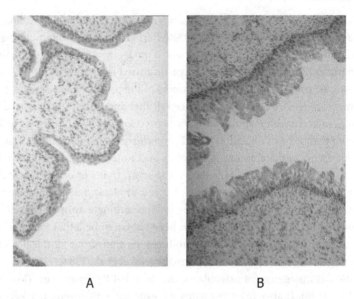

A   B

Figure 2.5. Proliferation of the bladder epithelium of pigs infected with toxigenic
P. multocida. A, normal bladder epithelium; B, epithelium from pig infected with
toxigenic *P. multocida*.

mostly non-toxigenic. However, those from respiratory infections likely to
have been acquired from pigs, for example from farm workers, are sometimes
toxigenic (Donnio et al., 1991). Their long-term carriage could have serious
implications.

CONCLUSIONS

The *Pasteurella multocida* toxin is a highly unusual toxin that is mitogenic
for many cell types. Most bacterial toxins are associated with acute diseases
in which cells are killed or their function is severely impaired. Although
PMT has long been associated with the pig disease atrophic rhinitis, chronic
exposure to the toxin may also have hidden consequences as its effects on
cell signalling show many similarities with those of tumour promoters. Very
few studies have looked at this possibility, although it is now widely accepted
that bacteria can cause cancer; in particular, *Helicobacter pylori* is linked with
gastric cancer (see Chapter 8). This is an important area for future work on
this toxin.

PMT is an activator of G_q-mediated signalling pathways, and possibly
also of the G_{12} class of G-proteins. It is the first toxin known to be an agonist

for G_q. Toxins that specifically activate or inactivate other heterotrimeric G-proteins, for example cholera toxin and pertussis toxin, have been widely used as reagents to help unravel cell-signalling pathways. It is anticipated that PMT will also become an increasingly valuable tool for cell biology research, particularly after its molecular mode of action is fully understood.

ACKNOWLEDGEMENT

I thank the Wellcome Trust for its support.

REFERENCES

Adams J W, Sakata Y, Davis M G, Sah V P, Wang Y, Liggett S B, Chien K R, Brown J H, and Dorn G W (1998). Enhanced $G\alpha_q$ signalling: A common pathway mediates cardiac hypertrophy and apoptotic heart failure. *Proc. Natl. Acad. Sci. USA*, **95**, 10140–10145.

Baldwin M R, Lakey J H, and Lax A J (2004). Identification and characterisation of the *Pasteurella multocida* toxin translocation domain. *Mol. Microbiol* (in press).

Baldwin M R, Pullinger G D, and Lax A J (2003). *Pasteurella multocida* toxin facilitates inositol phosphate formation by bombesin through tyrosine phosphorylation of $G\alpha_q$. *J. Biol. Chem.*, **278**, 32719–32725.

Busch C, Orth J, and Aktories K (2001). Biological activity of a C-terminal fragment of *Pasteurella multocida* toxin. *Infect. Immun.*, **69**, 3628–3634.

Buys W E C M, Smith H E, Kamps A M I E, Kamp E M, and Smits M A (1990). Sequence of the dermonecrotic toxin of *Pasteurella multocida* ssp. multocida. *Nucleic. Acids Res.*, **18**, 2815–2816.

Chanter N, Rutter J M, and Mackenzie A (1986). Partial purification of an osteolytic toxin from *Pasteurella multocida*. *J. Gen. Microbiol.*, **132**, 1089–1097.

Chardin P, Boquet P, Modaule P, Popoff M R, Rubin E J, and Gill D M (1989). The mammalian G protein RhoC is ADP-ribosylated by *Clostridium botulinum* exoenzyme C3 and affects actin microfilaments in Vero cells. *EMBO J.*, **8**, 1087–1092.

Donnio P Y, Avril J L, Andre P M, and Vaucel J (1991). Dermonecrotic toxin production by strains of *Pasteurella multocida* isolated from man. *J. Med. Microbiol.*, **34**, 333–337.

Dudet L I, Chailler P, Dubreuil J D, and Martineau-Doize B (1996). *Pasteurella multocida* toxin stimulates mitogenesis and cytoskeleton reorganization in Swiss 3T3 fibroblasts. *J. Cell Physiol.*, **168**, 173–182.

Dyer N W, Haynes J S, Ackermann M R, and Rimler R (1998). Morphological effects of *Pasteurella multocida* type-D dermonecrotoxin on rat osteosarcoma cells in a nude mouse model. *J. Comp. Pathol.*, **119**, 149–158.

Essler M, Hermann K, Amano M, Kaibuchi K, Heesemann J, Weber P C, and Aepfelbacher M (1998). *Pasteurella multocida* toxin increases endothelial permeability via Rho kinase and myosin light chain phosphatase. *J. Immunol.*, **161**, 5640–5646.

Fukuhara S, Murga C, Zohar M, Igishi T, and Gutkind J S (1999). A novel PDZ domain containing guanine nucleotide exchange factor links heterotrimeric G proteins to Rho. *J. Biol. Chem.*, **274**, 5868–5879.

Gilman A G (1987). G proteins: Transducers of receptor-generated signals. *Annu. Rev. Biochem.*, **56**, 615–649.

Gutkind J S (1998). The pathways connecting G protein-coupled receptors to the nucleus through divergent mitogen-activated protein kinase cascades. *J. Biol. Chem.*, **273**, 1839–1842.

Hart M J, Jiang X, Kozasa T, Roscoe W, Singer W D, Gilman A G, Sternweis P C, and Bollag G (1998). Direct stimulation of the guanine nucleotide exchange activity of p115 RhoGEF by $G\alpha_{13}$. *Science*, **280**, 2112–2114.

Higgins T E, Murphy A C, Staddon J M, Lax A J, and Rozengurt E (1992). *Pasteurella multocida* toxin is a potent inducer of anchorage-independent cell growth. *Proc. Natl. Acad. Sci. USA*, **89**, 4240–4244.

Hoskins I C, Thomas L H, and Lax A J (1997). Nasal infection with *Pasteurella multocida* causes proliferation of bladder epithelium in gnotobiotic pigs. *Vet. Rec.*, **140**, 22.

James S R and Downes C P (1997). Structural and mechanistic features of phospholipases C: Effectors of inositol phospholipid-mediated signal transduction. *Cell. Signal.*, **5**, 329–336.

Kamps A M I E, Kamp E M, and Smits M A (1990). Cloning and expression of the dermonecrotic toxin gene of *Pasteurella multocida* ssp. multocida in *Escherichia coli*. *FEMS Microbiol. Lett.*, **67**, 187–190.

Kimura K, Ito M, Amano M, Chihara K, Fukata Y, Nakafuku M, Yamamori B, Feng J, Nakano T, Okawa K, Iwamatsu A, and Kaibuchi K (1996). Regulation of myosin phosphatase by Rho and Rho-associated kinase (Rho-kinase). *Science*, **273**, 245–248.

Kozasa T, Jiang X, Hart M J, Sternweis P M, Singer W D, Gilman A G, Bollag G, and Sternweis P C (1998). p115 RhoGEF, a GTPase activating protein for $G\alpha_{12}$ and $G\alpha_{13}$. *Science*, **280**, 2109–2111.

Lacerda H M, Lax A J, and Rozengurt E (1996). *Pasteurella multocida* toxin, a potent intracellularly acting mitogen, induces p125[FAK] and paxillin tyrosine phosphorylation, actin stress fiber formation, and focal contact assembly in Swiss 3T3 cells. *J. Biol. Chem.*, **271**, 439–445.

Lacerda H M, Pullinger G D, Lax A J, and Rozengurt E. (1997). Cytotoxic necrotizing factor 1 from *Escherichia coli* and dermonecrotic toxin from *Bordetella bronchiseptica* induce p21^{rho}-dependent tyrosine phosphorylation of focal adhesion kinase and paxillin in Swiss 3T3 cells. *J. Biol. Chem.*, **272**, 9587–9596.

Lax A J and Chanter N (1990). Cloning of the toxin gene from *Pasteurella multocida* and its role in atrophic rhinitis. *J. Gen. Microbiol.*, **136**, 81–87.

Lax A J, Chanter N, Pullinger G D, Higgins T, Staddon J M, and Rozengurt E (1990). Sequence analysis of the potent mitogenic toxin of *Pasteurella multocida*. *FEBS Lett.*, **277**, 59–64.

Lemichez E, Flatau G, Bruzzone M, Boquet P, and Gauthier M (1997). Molecular localization of the *Escherichia coli* cytotoxic necrotizing factor CNF1 cell-binding and catalytic domains. *Mol. Microbiol.*, **24**, 1061–1070.

Mullan P B and Lax A J (1996). *Pasteurella multocida* toxin is a mitogen for bone cells in primary culture. *Infect. Immun.*, **64**, 959–965.

Murphy A C and Rozengurt E (1992). *Pasteurella multocida* toxin selectively facilitates phosphatidylinositol 4,5-bisphosphate hydrolysis by bombesin, vasopressin and endothelin. Requirement for a functional G protein. *J. Biol. Chem.*, **267**, 25296–25303.

Nakai T, Sawata A, Tsuji M, Samejima Y, and Kume K (1984). Purification of dermonecrotic toxin from a sonic extract of *Pasteurella multocida* SP-72 serotype D. *Infect. Immun.*, **46**, 429–434.

Neer E J (1995). Heterotrimeric G proteins: Organizers of transmembrane signals. *Cell*, **80**, 249–257.

Olsnes S, van Deurs B, and Sandvig K (1993). Protein toxins acting on intracellular targets: Cellular uptake and translocation to the cytosol. *Med. Microbiol. Immun.*, **182**, 51–61.

Orth J H C, Blocker D, and Aktories K (2003). His1205 and His1223 are essential for the activity of the mitogenic *Pasteurella multocida* toxin. *Biochemistry*, **42**, 4971–4977.

Parsons J T and Parsons S J (1997). Src family protein tyrosine kinases: Cooperating with growth factor and adhesion signalling pathways. *Curr. Opin. Cell Biol.*, **9**, 187–192.

Pedersen K B and Barfod K (1981). The aetiological significance of *Bordetella bronchiseptica* and *Pasteurella multocida* in atrophic rhinitis of swine. *Nord. Vet. Med.*, **33**, 513–522.

Petersen S K (1990). The complete nucleotide sequence of the *Pasteurella multocida* toxin gene and evidence for a transcriptional repressor, TxaR. *Mol. Microbiol.*, **4**, 821–830.

Petersen S K and Foged N T (1989). Cloning and expression of the *Pasteurella multocida* toxin gene, toxA, in *Escherichia coli*. *Infect. Immun.*, **57**, 3907–3913.

Petersen S K, Foged N T, Bording A, Nielsen J P, Riemann H K, and Frandsen P L (1991). Recombinant derivatives of *Pasteurella multocida* toxin: Candidates for a vaccine against progressive atrophic rhinitis. *Infect. Immun.*, **59**, 1387–1393.

Pettit R K, Ackermann M R, and Rimler R B (1993). Receptor-mediated binding of *Pasteurella multocida* dermonecrotic toxin to canine osteosarcoma and monkey kidney (vero) cells. *Lab. Invest.*, **69**, 94–100.

Pullinger G D, Bevir T, and Lax A J (2004). The *Pasteurella multocida* toxin is encoded within a lysogenic bacteriophage. *Mol. Microbiol.*, **51**, 255–269.

Pullinger G D, Sowdhamini R, and Lax A J (2001). Localization of functional domains of the mitogenic toxin of *Pasteurella multocida*. *Infect. Immun.*, **69**, 7839–7850.

Rhee S G (2001). Regulation of phosphoinositide-specific phospholipase C. *Annu. Rev. Biochem.*, **70**, 281–312.

Ridley A J and Hall A (1992). The small GTP-binding protein Rho regulates the assembly of focal adhesions and actin stress fibres in response to growth factors. *Cell*, **70**, 389–399.

Rozengurt E, Higgins T, Chanter N, Lax A J, and Staddon J M (1990). *Pasteurella multocida* toxin: Potent mitogen for cultured fibroblasts. *Proc. Natl. Acad. Sci. USA*, **87**, 123–127.

Rutter J M, Francis L M A, and Sansom B F (1982). Virulence of *Bordetella bronchiseptica* from pigs with or without atrophic rhinitis. *J. Med. Microbiol.*, **15**, 105–116.

Rutter J M and Luther P D (1984). Cell culture assay for toxigenic *Pasteurella multocida* from atrophic rhinitis of pigs. *Vet. Rec.*, **114**, 393–396.

Rutter J M and Mackenzie A (1984). Pathogenesis of atrophic rhinitis in pigs: A new perspective. *Vet. Rec.*, **114**, 89–90.

Rutter J M and Rojas X (1982). Atrophic rhinitis in gnotobiotic piglets: Differences in the pathogenicity of *Pasteurella multocida* in combined infections with *Bordetella bronchiseptica*. *Vet. Rec.*, **110**, 531–535.

Sabri A, Wilson B A, and Steinberg S F (2002). Dual actions of the $G\alpha_q$ agonist *Pasteurella multocida* toxin to promote cardiomyocyte hypertrophy and enhance apoptosis susceptibility. *Circ. Res.*, **90**, 850–857.

Sasaki T and Takai Y (1998). The Rho small G protein family-RhoGDI system as a temporal and spatial determinant for cytoskeletal control. *Biochem. Biophys. Res. Commun.*, **245**, 641–645.

Seo B, Choy E W, Maudsley S, Miller W E, Wilson B A, and Luttrell L M (2000). *Pasteurella multocida* toxin stimulates mitogen-activated protein kinase via $G_{q/11}$-dependent transactivation of the epidermal growth factor receptor. *J. Biol. Chem.*, **275**, 2239–2245.

Smyth M G, Pickersgill R W, and Lax A J (1995). The potent mitogen *Pasteurella multocida* toxin is highly resistant to proteolysis but becomes susceptible at lysosomal pH. *FEBS Lett.*, **360**, 62–66.

Smyth M G, Sumner I G, and Lax A J (1999). Reduced pH causes structural changes in the potent mitogenic toxin of *Pasteurella multocida*. *FEMS Microbiol. Lett.*, **180**, 15–20.

Sprang S R (1997). G protein mechanisms: Insights from structural analysis. *Annu. Rev. Biochem.*, **66**, 639–678.

Staddon J M, Barker C J, Murphy A C, Chanter N, Lax A J, Michell R H, and Rozengurt E (1991a). *Pasteurella multocida* toxin, a potent mitogen, increases inositol 1,4,5-trisphosphate and mobilizes Ca^{2+} in Swiss 3T3 cells. *J. Biol. Chem.*, **266**, 4840–4847.

Staddon J M, Bouzyk M M, and Rozengurt E (1991b). A novel approach to detect toxin-catalyzed ADP-ribosylation in intact cells: Its use to study the action of *Pasteurella multocida* toxin. *J. Cell Biol.*, **115**, 949–958.

Staddon J M, Chanter N, Lax A J, Higgins T E, and Rozengurt E (1990). Pasteurella multocida toxin, a potent mitogen, stimulates protein kinase C-dependent and -independent protein phosphorylation in Swiss 3T3 cells. *J. Biol. Chem.*, **265**, 11841–11848.

Switzer W P and Farrington D O (1975). Infectious atrophic rhinitis. In *Diseases of Swine*, ed H. W. Dunne and A. D. Leman, 4th edn, pp. 687–711. Ames: Iowa State University Press.

Thomas W, Pullinger G D, Lax A J, and Rozengurt E (2001). Escherichia coli cytotoxic necrotizing factor and *Pasteurella multocida* toxin induce focal adhesion kinase autophosphorylation and Src association. *Infect. Immun.*, **69**, 5931–5935.

Varmus H E (1984). The molecular genetics of cellular oncogenes. *Annu. Rev. Genet.*, **18**, 553–612.

Ward P N, Higgins T E, Murphy A C, Mullan P B, Rozengurt E, and Lax A J (1994). Mutation of a putative ADP-ribosylation motif in the *Pasteurella multocida* toxin does not affect mitogenic activity. *FEBS Lett.*, **342**, 81–84.

Ward P N, Miles A J, Sumner I G, Thomas L H, and Lax A J (1998). Activity of the mitogenic *Pasteurella multocida* toxin requires an essential C-terminal residue. *Infect. Immun.*, **66**, 5636–5642.

Wilson B A, Aminova L R, Ponferrada V G, and Ho M (2000). Differential modulation and subsequent blockade of mitogenic signalling and cell cycle progression by *Pasteurella multocida* toxin. *Infect. Immun.*, **68**, 4531–4538.

Wilson B A, Ponferrada V G, Vallance J E, and Ho M (1999). Localization of the intracellular activity domain of *Pasteurella multocida* toxin to the N terminus. *Infect. Immun.*, **67**, 80–87.

Wilson B A, Zhu X, Ho M, and Lu L (1997). Pasteurella multocida toxin activates the inositol trisphosphate signaling pathway in *Xenopus* oocytes via $G_q\alpha$-coupled phospholipase C-β1. *J. Biol. Chem.*, **272**, 1268–1275.

Zywietz A, Gohla A, Schmelz M, Schultz G, and Offermanns S (2001). Pleiotropic effects of *Pasteurella multocida* toxin are mediated by G_q-dependent and -independent mechanisms. Involvement of G_q but not G_{11}. *J. Biol. Chem.*, **276**, 3840–3845.

CHAPTER 3

Rho-activating toxins and growth regulation

Gudula Schmidt and Klaus Aktories

Low molecular mass GTPases of the Rho family are master regulators of the actin cytoskeleton and are involved in various signal transduction processes. They regulate the actin cytoskeleton and control transcriptional activation, cell cycle progression, cell transformation, and apoptosis. Rho GTPases are the eukaryotic targets of a variety of bacterial protein toxins that either inhibit or activate the Rho proteins. The cytotoxic necrotizing factors CNF1 and 2 from *Escherichia coli* and the dermonecrotic toxin DNT from *Bordetella* species are Rho-activating toxins. Whereas CNFs cause deamidation of Rho proteins, DNT deamidates and/or transglutaminates the GTPases by attachment of polyamines. Both deamidation and transglutamination cause persistent activation of Rho GTPases and subsequent changes in processes governed by the GTPases.

RHO GTPASES

Rho GTPases belong to the Ras superfamily of low molecular mass GTPases and are molecular switches in various signalling pathways. Like all members of GTP-binding proteins, they are active in the GTP bound form and inactive with GDP bound. Activation occurs by GDP/GTP exchange catalysed by guanine nucleotide exchange factors (GEFs) and is inhibited by binding of guanine nucleotide dissociation inhibitors (GDIs). Inactivation by hydrolysis of the bound GTP is catalysed by GTPase activating proteins (GAPs) (Symons and Settleman, 2000).

Rho GTPases are ubiquitously expressed. The members (>15) of the Rho family of GTPases, including RhoA, B, C, D, E, and G, Cdc42, Rac1, 2, and 3, share more than 50% sequence identity. The GTPases have been described as important regulators of the actin cytoskeleton. RhoA regulates the formation

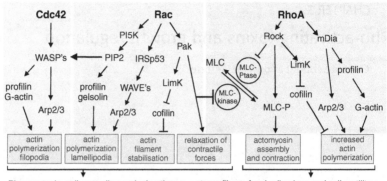

Phagocytosis, cell spreading and migration   stress fibers, focal adhesions and cell motility

Figure 3.1. Rho GTPases regulate the actin architecture by different pathways. Cdc42, Rac, and RhoA activate the Arp2/3 complex via different effector molecules, which are protein kinases (e.g., PI5K, PAK, ROCK) or scaffolding molecules (e.g., WASP, IRS p53, mDia). These effector molecules also regulate other signalling pathways, leading to actin polymerisation or depolymerisation and, thus, control cellular morphology, movement, and other actin-dependent processes.

of actin stress fibres – bundles of actin filaments associated with myosin filaments and other proteins. Cdc42 is known to induce formation of filopodia, which are small actin protrusions. Similar to stress fibres, filopodia are associated with focal contacts, which link the cytoskeleton with the extracellular matrix and the substratum. It has been suggested that filopodia are sensory elements more than the driving force for migrating cells. The third well-characterised member of the Rho family, Rac, is involved in the formation of lamellipodia and membrane ruffles. These actin-containing structures are necessary for phagocytosis and cell movement. By regulating the actin cytoskeleton, Rho-GTPases are involved in migration, phagocytosis, endo- and exocytosis, and cell–cell and cell–matrix contact (Nobes and Hall, 1999).

Furthermore, the family of Rho GTPases is involved in the regulation of transcriptional activation, cell cycle progression, cell transformation, and apoptosis. Many Rho GTPase effectors involved in these processes have been identified, including protein- and lipid kinases, phospholipase D, scaffolding proteins, and numerous adaptor proteins (Bishop and Hall, 2000).

The pathways leading to actin-based motility are the best characterised (Figure 3.1). RhoA-dependent reorganisation of the actin cytoskeleton is mediated by the cooperative action of Rho kinase (p160/ROCK) (Rho-associated kinase) and mDia (diaphanous) proteins. mDia proteins interact with actin via their C-terminal FH2 domain (Alberts, 2001) and regulate the profilin–G-actin interaction via their FH1 domain, thereby leading to actin polymerisation. Rho kinase, on the other hand, increases the myosin light chain (MLC) phosphorylation by inhibiting the MLC phosphatase and, thereby, facilitates the assembly of actin and myosin (Essler et al., 1998). Moreover, Rho-kinase, and also protein kinase C (PKC), phosphorylates calponin and, thereby, reduces calponin–actin interaction (Kaneko et al., 2000). Binding of calponin to actin has an inhibitory effect on the ATPase activity of myosin (for review see Fukata et al., 2001). In some cells, Rac antagonises the action of RhoA by inhibiting the MLC kinase, myosin phosphorylation, and the actomyosin assembly. Inactivation of the MLC kinase by Rac involves PAK, an effector of Rac and Cdc42. Besides inhibiting contractile forces, PAK leads to the stabilisation of actin filaments by activating LIM kinase, which inhibits cofilin by phosphorylation. Cofilin severs actin filaments and additionally increases the rate of monomer release at the pointed ends of actin filaments. However, Rho kinase also activates Lim kinase to act in this case synergistically with Rac signalling pathways. This effect might be of importance for RhoA-induced actin polymerisation. Activated Rac and Cdc42 stimulate actin nucleation by the Arp2/3 complex. This involves WASP (Wiskott-Aldrich syndrome protein) family proteins. Whereas WASP and the ubiquitously expressed N-WASP proteins are activated by Cdc42, WAVE (WASP verproline homologous protein) is regulated by Rac.

RHO GTPASES AS REGULATORS OF PROLIFERATION

The precise pathways whereby Rho proteins impinge on the regulation of proliferation are not particularly clear. In mammalian cells, proliferation is cooperatively controlled by soluble growth factors and cell adhesion to the extracellular matrix. Signalling pathways induced by both stimuli cross-talk on several regulatory levels and are regulated by various Rho GTPases. These include regulation of integrin signalling, control of MAP kinase cascades subsequent to tyrosine kinase receptor activation, and activation of PI3kinase-dependent pathways (Assoian and Schwartz, 2001; Danen and Yamada, 2001). For example, mitogenic signals from soluble growth factors are transduced by the canonical Ras-Raf-MEK-ERK cascade with subsequent transcriptional activation. Integrin signalling converges via focal adhesion kinase (p125^{FAK}), paxillin, and src kinases with MAP kinase cascade regulation. The functional consequences are transcriptional activation (e.g., increase in cyclin D1 transcription) and cell cycle progression, eventually resulting in cell growth.

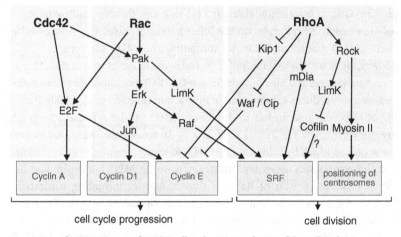

cell cycle progression   cell division

Figure 3.2. Rho GTPases regulate the cell cycle. For regulation of the cell cycle progression Cdc42, Rac, and RhoA control different members of cell cycle regulatory proteins of the cyclin family by activating (E2F, PAK) or inhibiting (Kip1, Waf/Cip) effector molecules. Cyclin A and D are targets for Cdc42 and Rac signalling, whereas RhoA activity primarily inhibits the CDK inhibitors Kip and Waf/Cip. The GTPases also control cell division by activating SRF-dependent transcription and regulating cytokinesis.

Rac/Cdc42 is known to activate the Jun kinase signalling cascade with c-Jun transcriptional activation (Figure 3.2) (Minden et al., 1995). However, a major role of Rac/Cdc42 is the control of cyclin D1 transcription. Activated mutants of Rac induce cyclin D1 (Page et al., 1999; Westwick et al., 1997). Moreover, constitutively activated Rac and Cdc42, but not RhoA, induce E2F activity and anchorage-independent induction of cyclin A (Philips et al., 2000). On the other hand, negative Rac mutants inhibit the cyclin D1 induction by oncogenic Ras (Gille and Downward, 1999). The Rac/Cdc42 effector Pak kinase is involved in integrin-induced Erk activation and stimulates MEK1 and Raf (Chaudhary et al., 2000). This process additionally involves PI3-kinase.

The major role of Rho in G1 progression appears to be different (Figure 3.2). Besides an effect of RhoA on cyclin D1 protein accumulation (Danen et al., 2000), the GTPase was shown to be involved in downregulation of inhibitors of cyclin-dependent kinases. Mitogen-induced Erk activation induces the CDK inhibitor p21[Waf/Cip] (Bottazzi et al., 1999) which is Rho dependent. Similarly Rho is essential to prevent p21[Waf/Cip] induction by oncogenic Ras (Olson et al., 1998). Furthermore, Rho appears to be involved in degradation of the CDK inhibitor p27[kip1] (Hirai et al., 1997). In line with these findings are studies showing that Rho GTPases are essential for cell transformation and possess at least some cell transformation activity. The role of Rho GTPases in transformation is more obviously appreciated by the observation that many GEFs are well-established products of oncogenes. This category includes Vav, (Olson et al., 1996), Lbc (Zheng et al., 1995), Dbl (Hart et al., 1991), TIAM-1 (Michiels et al., 1995), and many others (Jaffe and Hall, 2002). Accordingly, it was shown that mice, which are deficient in the Rac GEF TIAM, are resistant to Ras-induced skin tumours (Malliri et al., 2002).

Recently, an important connection between the actin cytoskeleton and transcriptional activation was described. It was shown that LIM Kinase and Diaphanous cooperate to regulate serum responsive factor and actin dynamics (Geneste et al., 2002). It has been known for many years that a dynamic actin cytoskeleton is needed for the cleavage of a dividing cell into two daughter cells. Moreover, it has been shown in many different cell systems that Rho GTPases are involved in cell division. Several Rho effectors, including Rho kinase (ROCK) and citron kinase, are localised at cleavage furrows (Chevrier et al., 2002). Moreover, Myosin II, which is phosphorylated by Rho kinase, is an essential motor for cytokinesis (Matsumura et al., 2001). Rho kinase has been identified as a component of the centrosome. It is required for positioning of the centrosomes, which play a role in cell division as well as in cell motility.

RHO PROTEINS AS TARGETS OF BACTERIAL TOXINS

During the last few years, it has been recognised that Rho proteins are major eukaryotic targets for various bacterial protein toxins. Some toxins block the functions of Rho GTPases by covalent modification. For example, C3-like toxins from *C. botulinum*, *C. limosum*, and *S. aureus*, which share 30 to 70% aminoacid sequence identity, ADP-ribosylate small GTPases of the Rho family (e.g., at Asn41 of RhoA [Sekine et al., 1989]) and inactivate them. The prototype of these small toxins (23–30 kDa) is the *Clostridium botulinum* C3 toxin, which ADP ribosylates RhoA, B, and C (Aktories et al., 1987; Wilde and Aktories, 2001).

It was assumed that Rho function is blocked due to sterical hinderance of the GTPase-effector interaction, because the modified residue is located close to the effector region. However, recent studies indicate that ADP-ribosylated Rho is still able to interact with at least some effectors (Genth et al., 2003b). However, the rate of activation of Rho by exchange factors (e.g., Lbc) is diminished by ADP-ribosylation (Genth et al., 2003b), and it was suggested that ADP-ribosylation prevents the conformational change, occurring subsequently with GDP/GTP exchange (Genth et al., 2003a). Moreover, ADP-ribosylated Rho is released from membranes and forms a tight complex with GDI, a guanine nucleotide dissociation inhibitor, which keeps Rho in its inactive GDP-bound form in the cytosol (Genth et al., 2003a).

C3-like exoenzymes consist of only the enzyme domain and lack a specific cell membrane–binding and translocation unit. The uptake of C3-like toxins into target cells and their potential roles as virulence factors are not well understood. Two explanations are possible. First, it was suggested that the enzymes (at least those produced by *S. aureus*) are directly released into the cytosol from bacteria, which are capable of invading eukaryotic target cells (Wilde et al., 2001). Second, uptake of C3 exoenzymes might depend on the presence of bacterial pore-forming toxins, which facilitate translocation. C3 toxins are widely used to inactivate RhoA. For the use of the RhoA-specific C3 toxins as pharmacological tools, toxin chimeras, consisting of C3 and the binding and translocation domain of "complete" toxins (e.g., *C. botulinum* C2 toxin), have been constructed (Barth et al., 2002).

Large clostridial cytotoxins comprise a second family of Rho protein inactivating toxins. These toxins modify the GTPases by glucosylation (Busch and Aktories 2000; Just et al., 1995; Just et al., 2000). Members of this toxin family are *C. difficile* toxins A and B, including various isoforms, the lethal and the haemorrhagic toxins from *C. sordellii*, and the alpha toxin from *C. novyi*. All these toxins are single-chain proteins with molecular masses of 250 to 308 kDa and encompass a catalytic domain and a specific binding and translocation domain. The substrate specificity of large clostridial toxins is broader than that of C3-like toxins. For example, *C. difficile* toxins A and B glucosylate many GTPases of the Rho family, including Rho A, B and C, Rac and Cdc42. *C. sordellii* lethal toxin possesses a different substrate specificity and modifies Rac but not RhoA. In addition, Ras subfamily proteins (e.g., Ras, Ral, and Rap) are glucosylated (Just et al., 1996). *C. novyi* toxin, which shares the substrate specificity of toxin B, is an *O*-GlcNAc transferase.

All these transferases modify a highly conserved threonine residue (e.g., Thr37 in RhoA) in the switch 1 region of the GTPases, which is involved in Mg^{2+} and nucleotide binding. Modification of this threonine residue by mono-*O*-glucosylation has the following effects: (1) causes inhibition of the interaction of GTPases with their effectors; (2) increases membrane binding; (3) blocks activation by exchange factors; and (4) inhibits intrinsic and GAP-stimulated GTPase activity (Genth et al., 1999; Sehr et al., 1998). The toxins are taken up from an acidic endosomal compartment and glucosylate RhoA, Rac, and Cdc42 in the cytosol (Barth et al., 2001). Glucosylating toxins are also widely used as tools to study the functions of GTPases.

Inactivating Rho GTPases is not the only way to influence signal transduction pathways of mammalian host cells. Rho proteins are also activated due to covalent modification catalysed by bacterial protein toxins like the cytotoxic necrotizing factors CNF1 and CNF2 from *E. coli* and the dermonecrotic toxin DNT from *Bordetella* species (described below). Recent studies indicate that Rho proteins are not exclusively covalently modified by bacterial toxins. Some bacterial effectors, like the *Salmonella* SopEs and SptP, modulate the activity of Rho GTPases by acting as regulatory proteins with GAP (SptP) or GEF (SopEs) functions (see Chapter 6).

CNFs Activate Rho GTPASES

In 1983, Caprioli and co-workers isolated a toxin from an *Escherichia coli* obtained from enteritis-affected children. Because of the necrotising effects on rabbit skin they called the toxin CNF (cytotoxic necrotizing factor) (Caprioli et al., 1983). Besides the skin necrotising action, CNF turned out to be lethal for animals after i.p. injection. The lethal dose ($LD_{50,\,mice}$) was estimated to be about 20 ng of purified material (de Rycke et al., 1997). Studies with cultured cells revealed typical morphological changes, cell body enlargement, stress fibre formation, and multinucleation. Subsequently, CNF1 and later the homologue CNF2, first named Vir cytotoxin (Oswald et al., 1989), was found in various pathogenic *E. coli* strains isolated from animals (e.g., piglets and calves) (de Rycke et al., 1987; de Rycke et al., 1990) and man.

Structure and Up-Take of CNFs

CNF1 and CNF2 are closely related toxins, sharing more than 90% identity in their amino acid sequences (Oswald et al., 1994). Whereas CNF1 is chromosomally encoded, CNF2 is encoded by transmissible plasmids (Oswald and de Rycke, 1990). Both toxins are single-chain proteins with molecular masses of about 115 kDa. They are constructed like AB toxins, with the cell-binding and catalytic domains located at the N terminus (amino acids 53 to 190, cell-binding domain) and C terminus (amino acids 720 to 1014, catalytic domain) of the toxin, respectively (Lemichez et al., 1997). The central part appears to be involved in membrane translocation. The receptor for the entry of the toxin into cells is still unknown. Uptake of CNF appears to occur by clathrin-dependent and -independent endocytosis, which is followed by cell entry from an acidic endosomal compartment (Contamin et al., 2000). Recently, two hydrophobic helices (aa350–412), which are separated by a short loop (aa 373–386), have been suggested to be involved in membrane insertion

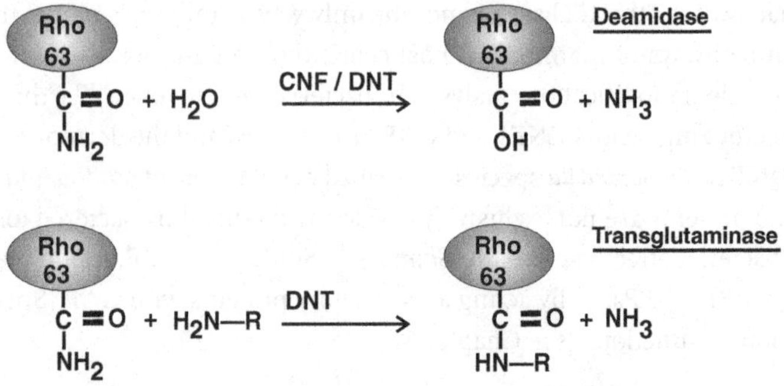

Figure 3.3. Molecular mechanism of Rho activation by CNF and DNT. Glutamine 63 of RhoA (Gln 61 of Rac and Cdc42) is essential for the hydrolysis of bound GTP. CNF and DNT deamidate this glutamine residue, creating glutamic acid, and the GTP hydrolysing activity of the GTPase is blocked. In the presence of primary amines, the toxins can polyaminate Rho at the same residue, thereby also blocking hydrolysis of GTP. Polyamination is the preferred activity of DNT whereas CNF is a better deamidase.

in a similar manner to the hairpin helices TH 8–9 of diphtheria toxin (Pei et al., 2001).

Mode of Action of CNFs

CNFs are cytotoxic for a wide variety of cells, including 3T3 fibroblasts, Chinese hamster ovary cells (CHO), Vero cells, HeLa cells, and cell lines of neuronal origin. The toxins lead to enlargement and flattening of culture cells in a time- and concentration-dependent manner. These changes are accompanied by transient and early formation of filopodia and membrane ruffles and a dense network of actin stress fibres, indicating that Rho proteins are involved in the action of these toxins. CNFs change the migration behaviour of Rho in SDS-PAGE (Oswald and de Rycke, 1990). This finding suggested a covalent modification of Rho GTPases by CNFs and allowed the elucidation of their mode of action. To this end, mass spectrometric analysis showed that CNF1 causes an increase in mass of RhoA by 1 Da. This change in mass is due to a deamidation at glutamine 63 of RhoA (see reaction scheme in Figure 3.3). Glutamine 63 in RhoA is essential for the GTP hydrolysing activity of the GTPase. Thus, GTP hydrolysis is blocked after treatment of Rho with CNF1.

Moreover, the stimulation of RhoA GTPase activity by GAP is blocked after CNF1/2 treatment and RhoA is held constitutively active. Similarly, Rac and Cdc42 are deamidated by CNF (Lerm et al., 1999b). Deamidation occurs at the equivalent amino acid residue glutamine 61. Although CNF is highly specific for Rho GTPases, recent studies show that even a small peptide covering the switch–II region of Rho GTPases is sufficient for substrate recognition by CNFs.

The crystal structure of the enzyme domain of CNF1 has been solved (Buetow et al., 2001), showing a novel protein fold. The structure confirmed the previous suggestions that CNF belongs to the catalytic triad family. In fact it was shown that a cysteine (Cys 866) and a histidine residue (His 881) are essential for enzyme activity (Schmidt et al., 1998). As identified from the crystal structure, the third "catalytic" residue appears to be a valine residue (Val 833), a finding which is rather unusual among catalytic triad enzymes (Buetow et al., 2001). Crystallisation of catalytic domains of various bacterial enzyme toxins (e.g., ExoS GAP domain [Würtele et al., 2001]), which share regulatory mammalian counterparts, indicates that the overall structure of the enzyme domain of bacterial toxins is not necessarily similar to their mammalian counterparts. The same is true for the catalytic domain of CNF1, which exhibits a different protein fold as compared with mammalian transglutaminases (Pedersen et al., 1994). A reason for the high specificity of CNF may be the existence of a deep cleft in the molecule with the catalytic cysteine at the bottom (Buetow et al., 2001). Recently, a potential role in substrate recognition has been described for three of nine loops located on the surface of the catalytic domain of CNF1 (Buetow and Ghosh, 2003). The structure of the CNF1 catalytic region contributes to the idea of a convergent evolution of the toxins and mammalian enzymes with the same mechanism, rather than gene transfer.

Cell Biological Effects

As already mentioned, the effects of CNF in cultured cells are characterised by major changes of actin structure, including stress fibres, lamellipodia, and filopodia. An increase in the formation of lamellipodia and membrane ruffles is prototypic for enhanced phagocytic and endocytic activity. Accordingly, activation of Rho GTPases by CNF induces phagocytic behaviour and macropinocytosis in mammalian cells (e.g., human epithelial cells), which are non-professional phagocytes (Falzano et al., 1993; Fiorentini et al., 2001). Quite early, it was found that CNF causes phosphorylation of paxillin and Fak kinase, which are known to be involved in nuclear signalling. This pathway is Rho dependent but does not involve the classical Map kinase pathway (Lacerda et al., 1997). One of the most striking effects observed with CNF is the formation of multinucleation (Oswald et al., 1989). The effect might be caused by blocking cell division without changes in nuclear cycling, or by increase in the rate of nuclear cycles within one cell division cycle (Denko et al. 1997). Recent data show that CNF2 uncouples S-phase from mitosis. Thus, it affects cytoplasmic division and removes the requirement for a complete mitosis before starting another S-phase. CNF increases the expression of the cyclooxygenase-2 gene in fibroblasts (Thomas et al., 2001). This finding is of importance because an increase in cyclooxygenase expression is observed in several tumours and it has been suggested that lipid mediators produced by the enzyme are responsible for tumour progression (Oshima et al., 1996).

Rac, which is also a substrate for CNF-induced deamidation, is involved in Jun kinase activation and, therefore, CNF1 causes activation of this kinase. Surprisingly, c-Jun kinase activity is only transiently increased after CNF treatment of cells, although the GTPases are constitutively activated (Lerm et al., 1999b). Recently, the reason for the transient activation was identified as degradation of CNF-activated Rac by a proteasome-dependent pathway (Lerm et al., 2002). Thus, it appears that the targeted cell is able to block the persistent activation of Rac induced by deamidation by rapid degradation. So far, it is not clear whether degradation of activated Rac is part of a general mechanism in mammalian cells to limit "overactivation" of GTPases or due to subtle structural changes induced by the deamidation.

DERMONECROTIC TOXIN (DNT) FROM *BORDETELLA*

A similar mechanism of Rho modification as described for CNF1 was reported for the CNF1-related dermonecrotic toxin (DNT) from *Bordetella* species (Horiguchi et al., 1997; Horiguchi 2001; Kashimoto et al., 1999; Schmidt et al., 1999).

DNT is produced by *Bordetella pertussis, B. parapertussis,* and *B. bronchiseptica*. Its name is derived from dermonecrotic effects caused by intradermal injection of the purified toxin (Horiguchi et al., 1989). DNT is a large, 160-kDa heat-labile protein, which shares significant sequence similarity with CNF (Figure 3.4). The sequence similarity is restricted to the C terminus of DNT and CNF, which in part harbours the catalytic activity of the deamidase, suggesting that both toxins share a similar molecular mechanism. In fact, DNT induces similar if not identical morphological changes (enlargement of cells, multinucleation, actin polymerisation) as CNF. Moreover, it was demonstrated that DNT causes a covalent modification of Rho, resulting in slightly slower migration of the GTPase in SDS-PAGE (Schmidt et al., 1999). In line with this notion, it was demonstrated that CNF1 and DNT induce deamidation of Rho at position Gln63. Similarly, as for CNF, the

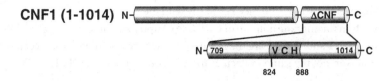

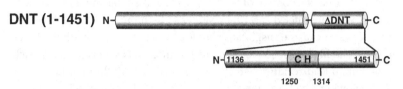

Figure 3.4. Homology between CNF1 and DNT. CNF and DNT consist of 1014 and 1451 amino acids, respectively. The toxins share homology within their catalytic domains (ΔCNF, aa 709 to 1014 and ΔDNT, aa 1136–1451) that are located at the C termini of the proteins. Highest sequence similarity is observed in a stretch of 64 amino acid residues, which covers residues 1250 through 1314 of DNT, showing ~45% sequence identity, whereas the amino acid sequence of the whole catalytic domain of DNT is only ~13% identical with the sequence of the catalytic domain of CNF1.

DNT targets are not only Rho but also Cdc42 and Rac. Both toxins were shown to catalyse the polyamination of glutamine 63/61 of Rho GTPases in the presence of rather high concentrations of amines. However, it appears that DNT prefers transglutamination, whereas CNF is primarily a deamidase (Figure 3.3) (Schmidt et al., 1999). Recently, putrescine, spermidine, and spermine have been identified as *in vivo* substrates for the transglutamination (Masuda et al. 2000; Schmidt et al. 2001). Lysine is a very good substrate for transglutamination by DNT at least *in vitro* (Schmidt et al., 2001).

STRUCTURE–FUNCTION RELATIONSHIP BETWEEN CNFs AND DNT

CNFs and DNT are both AB toxins, with a cell-binding domain located at the N-terminus and a C-terminal catalytic domain. Amino acids 1 to 531 of DNT blocked the intoxication of cells by full-length DNT, suggesting that this fragment retains the cell-binding domain of DNT (Kashimoto et al., 1999). More recently, the receptor-binding domain of DNT was mapped to amino acids 1 to 54 (Matsuzawa et al., 2002). Moreover, a Furin cleavage site within this binding domain was identified. Proteolytic processing of DNT by Furin seems to be necessary for translocation of the toxin across cellular membranes (Matsuzawa et al., 2004). It is of interest that the *Pasteurella multocida* toxin shares significant sequence similarity with the N terminus of DNT and with the transmembrane domain of CNF (see Chapter 2). Thus, it is suggested that both DNT and PMT share the same or the same type of membrane receptors, however, not the same molecular mechanism (Lemichez et al., 1997; Walker and Weiss, 1994).

As the catalytic domain of DNT is located at the C-terminus, a fragment of DNT was constructed covering residues 1136 through to 1451, which was fully active to cause transglutamination and deamidation of RhoGTPases *in vitro*. The highest sequence similarity with CNF is observed in a stretch of 64 amino acid residues, which covers residue 1250 through to 1314 of DNT, showing ~45% sequence identity, whereas the amino acid sequence of the minimal active fragment of DNT (residues 1136–1451) is only ~13% identical with the sequence of the minimal active fragment of CNF1 (residues 709–1014).

As mentioned above, the catalytic triad of CNF shares a typical cysteine and histidine residue with the catalytic centre of transglutaminases. DNT also possesses these conserved cysteine and histidine residues, which are Cys 1292 and His 1307. These catalytic residues share the same spacing as in CNF. However, biochemical studies also indicate differences in the enzyme activities of DNT and CNFs. For example, the minimal Rho sequence allowing deamidation or transglutamination by CNF1 is a peptide covering mainly the switch-II region (D59–D78) of RhoA. By contrast, DNT appears to need further interaction sites, and modifies exclusively the GDP-bound form of Rho GTPases (Lerm et al., 1999a). Therefore the enzyme substrate interaction for DNT appears to be more complex than for CNFs.

CNFs AND DNT AS VIRULENCE FACTORS

Initially, the role of CNF as a virulence factor was debated. However, recently, several studies have suggested that CNFs are important for *E. coli*–caused diseases. It has been shown that colonisation and tissue damage of the urinary tract of mice induced by CNF-producing *E. coli* strains are more severe than with CNF1-deficient isogenic strains (Rippere-Lampe et al., 2001b). The same group has reported that tissue damage of rat prostates with CNF1-producing uropathogenic *E. coli* strains is more extensive than after infection with isogenic CNF1-negative mutants (Rippere-Lampe et al., 2001a). Furthermore, it was reported that the toxin is involved in the *in vitro* invasion of brain microvascular endothelial cells by *E. coli* and contributes to the traversal of the blood–brain barrier in a meningitis animal model (Khan et al., 2002).

In contrast, no difference between CNF-producing and -deficient strains has been found when studying lung and serosal inflammation (Fournout et al., 2000). Nevertheless, taken together, one can say that CNF1 is involved in *E. coli* virulence and pathogenicity.

Another topic is of interest and concern. Rho-GTPases are increasingly recognised to be essential for proliferation, development of cancer, and metastasis. Because CNFs cause activation of Rho GTPases and mediate signalling leading to cell transformation, it is feasible that chronic carriers of CNF-producing *E. coli* might be challenged by a tumourigenic potential of the toxin (Lax and Thomas, 2002). This might be especially important for prostate carcinoma and colon tumours. In this respect, it appears important that CNFs also affect apoptotic processes. Although an increase in apoptosis by CNF was observed in uroepithelial 5637 cells (Mills et al., 2000), major anti-apoptotic effects of CNF were also reported (Fiorentini et al., 1998). The differences in outcome might be due to differences in cell types or toxins concentration.

The role of DNT as a virulence factor of *Bordetella bronchiseptica, B. pertussis*, and *B. parapertussis* is also not precisely defined. In fact, the toxin was described quite early as a virulence factor for whooping cough (Horiguchi, 2001). However, it is now accepted that DNT is not a major factor involved in this disease. By contrast, DNT is considered to be one of the major virulence factors in turbinate atrophy in pigs (Magyar et al., 1988). Its role in the pathogenesis of respiratory diseases has been shown by comparing isogenic DNT mutants with the corresponding wild-type strains in the efficiency of colonization of the respiratory tract of pigs. These studies showed that production of DNT by *B. bronchiseptica* is essential to induce lesions of turbinate atrophy and bronchopneumonia in pigs (Brockmeier et al., 2002). Accordingly, DNT was reported to affect osteoblastic MC3T3-E1 cells *in vitro* and to impair bone formation in neonatal rats (Horiguchi et al., 1995).

CONCLUSIONS

The deamidating and transglutaminating toxins CNFs and DNT, which activate Rho GTPases, have multiple effects on morphology, motility, proliferation, differentiation, and apoptosis of cells. Recent studies have shown that CNFs and DNT are major virulence factors in various infection models. Moreover, because Rho GTPases are crucial switches in signalling pathways responsible for cell transformation and metastasis, it is plausible (although still speculative) that they play a potential role in the pathogenesis of certain types of cancer.

Aktories K, Weller U, and Chhatwal G S (1987). *Clostridium botulinum* type C produces a novel ADP-ribosyltransferase distinct from botulinum C2 toxin. *FEBS Lett.*, **212**, 109–113.

Alberts A S (2001). Identification of a carboxyl-terminal diaphanous-related formin homology protein autoregulatorydomain. *J. Biol. Chem.*, **276**, 2824–2830.

Assoian R K and Schwartz M A (2001). Coordinate signaling by integrins and receptor tyrosine kinases in the regulation of G_1 phase cell-cycle progression. *Curr. Opin. Genet. Dev.*, **11**, 48–53.

Barth H, Pfeifer G, Hofmann F, Maier E, Benz R, and Aktories K (2001). Low pH-induced formation of ion channels by *Clostridium difficile* toxin B in target cells. *J. Biol. Chem.*, **276**, 10670–10676.

Barth H, Roebling R, Fritz M, and Aktories K (2002). The binary *Clostridium botulinum* C2 toxin as a protein delivery system. *J. Biol. Chem.*, **277**, 5074–5081.

Bishop A L and Hall A (2000). Rho GTPases and their effector proteins. *Biochem. J.*, **348**, 241–255.

Bottazzi M E, Zhu X, Bohmer R M, and Assoian R K (1999). Regulation of p21(cip1) expression by growth factors and the extracellular matrix reveals a role for transient ERK activity in G1 phase. *J. Cell Biol.*, **146**, 1255–1264.

Brockmeier S L, Register K B, Magyar T, Lax A J, Pullinger G D, and Kunkle R A (2002). Role of the dermonecrotic toxin of *Bordetella bronchiseptica* in the pathogenesis of respiratory disease in swine. *Infect. Immun.*, **70**, 481–490.

Buetow L, Flatau G, Chiu K, Boquet P, and Ghosh P (2001). Structure of the Rho-activating domain of *Escherichia coli* cytotoxic necrotizing factor 1. *Nat. Struct. Biol.*, **8**, 584–588.

Buetow L and Ghosh P (2003). Structural elements required for deamidation of RhoA by cytotoxic necrotizing factor 1. *Biochemistry US*, **42**, 12784–12791.

Busch C and Aktories K (2000). Microbial toxins and the glucosylation of Rho family GTPases. *Curr. Opin. Struc. Biol.*, **10**, 528–535.

Caprioli A, Falbo V, Roda L G, Ruggeri F M, and Zona C (1983). Partial purification and characterization of an *Escherichia coli* toxic factor that induces morphological cell alterations. *Infect. Immun.*, **39**, 1300–1306.

Chaudhary A, King W G, Mattaliano M D, Frost J A, Diaz B, Morrison D K, Cobb M H, Marshall M S, and Brugge J S (2000). Phosphatidylinositol 3-kinase regulates Raf1 through Pak phosphorylation of serine 338. *Curr. Biol.*, **10**, 551–554.

Chevrier V, Piel M, Collomb N, Saoudi Y, Frank R, Paintrand M, Narumiya S, Bornens M, and Job D (2002). The Rho-associated protein kinase p160ROCK is required for centrosome positioning. *J. Cell Biol.*, **157**, 807–817.

Contamin S, Galmiche A, Doye A, Flatau G, Benmerah A, and Boquet P (2000). The p21 Rho-activating toxin cytotoxic necrotizing factor 1 is endocytosed by a clathrin-independent mechanism and enters the cytosol by an acidic-dependent membrane translocation step. *Mol. Biol. Cell*, **11**, 1775–1787.

Danen E H J, Sonneveld P, Sonnenberg A, and Yamada K M (2000). Dual stimulation of Ras/mitogen-activated protein kinase and RhoA by cell adhesion to fibronectin supports growth factor-stimulated cell cycle progression. *J. Cell Biol.*, **151**, 1413–1422.

Danen E H J and Yamada K M (2001). Fibronectin, integrins, and growth control. *J. Cell. Physiol.*, **189**, 1–13.

de Rycke J, González E A, Blanco J, Oswald E, Blanco M, and Boivin R (1990). Evidence for two types of cytotoxic necrotizing factor in human and animal clinical isolates of *Escherichia coli*. *J. Clin. Microbiol.*, **28**, 694–699.

de Rycke J, Guillot J F, and Boivin R (1987). Cytotoxins in non-enterotoxigenic strains of *Escherichia coli* isolated from feces of diarrheic calves. *Vet. Microbiol.*, **15**, 137–150.

de Rycke J, Phan-Thanh L, and Bernard S (1997). Immunochemical identification and biological characterization of cytotoxic necrotizing factor from *Escherichia coli*. *J. Clin. Microbiol.*, **27**, 983–988.

Denko N, Langland R, Barton M, and Lieberman M A (1997). Uncoupling of S-phase and mitosis by recombinant cytotoxic necrotizing factor 2 (CNF2) *Exp. Cell Res.*, **234**, 132–138.

Essler M, Amano M, Kruse H-J, Kaibuchi K, Weber P C, and Aepfelbacher M (1998). Thrombin inactivates myosin light chain phosphatase via Rho and its target Rho kinase in human endothelial cells. *J. Biol. Chem.*, **273**, 21867–21874.

Falzano L, Fiorentini C, Donelli G, Michel E, Kocks C, Cossart P, Cabanié L, Oswald E, and Boquet P (1993). Induction of phagocytic behaviour in human epithelial cells by *Escherichia coli* cytotoxic necrotizing factor type 1. *Mol. Microbiol.*, **9**, 1247–1254.

Fiorentini C, Falzano L, Fabbri A, Stringaro A, Logozzi M, Travaglione S, Contamin S, Arancia G, Malorni W, and Fais S (2001). Activation of Rho GTPases by cytotoxic necrotizing factor 1 induces macropinocytosis and scavenging activity in epithelial cells. *Mol. Biol. Cell*, **12**, 2061–2073.

Fiorentini C, Matarrese P, Straface E, Falzano L, Donelli G, Boquet P, and Malorni W (1998). Rho-dependent cell spreading activated by *E. coli* cytotoxic necrotizing factor 1 hinders apoptosis in epithelial cells. *Cell Death Differ.*, **5**, 921–929.

Fournout S, Dozois C M, Odin M, Desautels C, Pérès S, Hérault F, Daigle F, Segafredo C, Laffitte J, Oswald E, Fairbrother J M, and Oswald I P (2000). Lack of a role of cytotoxic necrotizing factor 1 toxin from *Escherichia coli* in bacterial pathogenicity and host cytokine response in infected germfree piglets. *Infect. Immun.*, **68**, 839–847.

Fukata Y, Amano M, and Kaibuchi K (2001). Rho-Rho-kinase pathway in smooth muscle contraction and ccytoskeletal reorganization of non-muscle cells. *Trends Pharmacol. Sci.*, **22**, 32–39.

Geneste O, Copeland J W, and Treisman R (2002). LIM kinase and Diaphanous cooperate to regulate serum response factor and actin dynamics. *J. Cell Biol.*, **157**, 831–838.

Genth H, Aktories K, and Just I (1999). Monoglucosylation of RhoA at Threonine-37 blocks cytosol-membrane cycling. *J. Biol. Chem.*, **274**, 29050–29056.

Genth H, Gerhard R, Maeda A, Amano M, Kaibuchi K, Aktories K, and Just I (2003a). Entrapment of Rho ADP-ribosylated by *Clostridium botulinum* C3 exoenzyme in the Rho-GDI-1 complex. *J. Biol. Chem.*, **278**, 28523–28527.

Genth H, Schmidt M, Gerhard R, Aktories K, and Just I (2003b). Activation of phospholipase D1 by ADP-ribosylated RhoA. *Biochem. Biophys. Res. Commun.*, **302**, 127–132.

Gille H and Downward J (1999). Multiple ras effector pathways contribute to G(1) cell cycle progression. *J. Biol. Chem.*, **274**, 22033–22040.

Hart M J, Eva A, Evans T, Aaronson S A, and Cerione R A (1991). Catalysis of guanine nucleotide exchange on the CDC42Hs protein by the *dbl* oncogene product. *Nature*, **354**, 311–314.

Hirai A, Nakamura S, Noguchi Y, Yasuda T, Kitagawa M, Tatsuno I, Oeda T, Tahara K, Terano T, Narumiya S, Kohn L D, and Saito Y (1997). Geranylgeranylated Rho small GTPase(s) are essential for the degradation of p27[Kip1] and facilitate the progression from G_1 to S phase in growth-stimulated rat FRTL-5 cells. *J. Biol. Chem.*, **272**, 13–16.

Horiguchi Y (2001). *Escherichia coli* cytotoxic necrotizing factors and *Bordetella* dermonecrotic toxin: The dermonecrosis-inducing toxins activating Rho small GTPases. *Toxicon*, **39**, 1619–1627.

Horiguchi Y, Inoue N, Masuda M, Kashimoto T, Katahira J, Sugimoto N, and Matsuda M (1997). *Bordetella bronchiseptica* dermonecrotizing toxin induces reorganization of actin stress fibers through deamidation of Gln-63 of the GTP-binding protein Rho. *Proc. Natl. Acad. Sci. USA*, **94**, 11623–11626.

Horiguchi Y, Nakai T, and Kume K (1989). Purification and characterization of *Bordetella bronchiseptica* dermonecrotic toxin. *Microb. Pathog.*, **6**, 361–368.

Horiguchi Y, Okada T, Sugimoto N, Morikawa Y, Katahira J, and Matsuda M (1995). Effects of *Bordetella bronchiseptica* dermonecrotizing toxin on bone formation in calvaria of neonatal rats. *FEMS Microbiol. Lett.*, **12**, 29–32.

Jaffe A B and Hall A (2002). Rho GTPases in transformation and metastasis. *Adv Cancer Res*, **84**, 57–80.

Just I, Hofmann F, and Aktories K (2000). Molecular mechanisms of action of the large clostridial cytotoxins. In *Handbook of Experimental Pharmacology*, ed. K Aktories and I Just pp. 307–331. Springer Verlag, Berlin.

Just I, Selzer J, Hofmann F, Green G A, and Aktories K (1996). Inactivation of Ras by *Clostridium sordellii* lethal toxin-catalyzed glucosylation. *J. Biol. Chem.*, **271**, 10149–10153.

Just I, Selzer J, Wilm M, Von Eichel-Streiber C, Mann M, and Aktories K (1995). Glucosylation of Rho proteins by *Clostridium difficile* toxin B. *Nature*, **375**, 500–503.

Kaneko T, Amano M, Maeda A, Goto H, Takahashi K, Ito M, and Kaibuchi K (2000). Identification of calponin as a novel substrate of Rho-kinase. *Biochem. Biophys. Res. Commun.*, **273**, 110–116.

Kashimoto T, Katahira J, Cornejo W R, Masuda M, Fukuoh A, Matsuzawa T, Ohnishi T, and Horiguchi Y (1999). Identification of functional domains of *Bordetella* dermonecrotizing toxin. *Infect. Immun.*, **67**, 3727–3732.

Khan N A, Wang Y, Kim K J, Chung J W, Wass C A, and Kim K S (2002). Cytotoxic necrotizing factor-1 contributes to *Escherichia coli* K1 invasion of the central nervous system. *J. Biol. Chem.*, **277**, 15607–15612.

Lacerda H M, Pullinger G D, Lax A J, and Rozengurt E (1997). Cytotoxic necrotizing factor 1 from *Escherichia coli* and dermonecrotic toxin from *Bordetella bronchiseptica* induce p21[rho]-dependent tyrosine phosphorylation of focal adhesion kinase and paxillin in swiss 3T3 cells. *J. Biol. Chem.*, **272**, 9587–9596.

Lax A J and Thomas W (2002). How bacteria could cause cancer: One step at a time. *Trends Microbiol.*, **10**, 293–299.

Lemichez E, Flatau G, Bruzzone M, Boquet P, and Gauthier M (1997). Molecular localization of the *Escherichia coli* cytotoxic necrotizing factor CNF1 cell-binding and catalytic domains. *Mol. Microbiol.*, **24**, 1061–1070.

Lerm M, Pop M, Fritz G, Aktories K, and Schmidt G (2002). Proteasomal degradation of cytotoxic necrotizing factor 1-activated Rac. *Infect. Immun.*, **70**, 4053–4058.

Lerm M, Schmidt G, Goehring U-M, Schirmer J, and Aktories K (1999a). Identification of the region of Rho involved in substrate recognition by *Escherichia coli* cytotoxic necrotizing factor 1 (CNF1). *J. Biol. Chem.*, **274**, 28999–29004.

Lerm M, Selzer J, Hoffmeyer A, Rapp U R, Aktories K, and Schmidt G (1999b). Deamidation of Cdc42 and Rac by *Escherichia coli* cytotoxic necrotizing factor

1 (CNF1) – activation of c-Jun-N-terminal kinase in HeLa cells. *Infect. Immun.*, **67**, 496–503.

Magyar T, Chanter N, Lax A J, Rutter J M, and Hall G A (1988). The pathogenesis of turbinate atrophy in pigs caused by *Bordetella bronchiseptica*. *Vet. Microbiol.*, **18**, 135–146.

Malliri A, Van der Kammen R A, Clark K, van der Valk M, Michiels F, and Collard J G (2002). Mice deficient in the Rac activator Tiam1 are resistant to Ras-induced skin tumours. *Nature*, **417**, 867–871.

Masuda M, Betancourt L, Matsuzawa T, Kashimoto T, Takao T, Shimonishi Y, and Horiguchi Y (2000). Activation of Rho through a cross-link with polyamines catalyzed by *Bordetella* dermonecrotizing toxin. *EMBO J.*, **19**, 521–530.

Matsumura F, Totsukawa G, Yamakita Y, and Yamashiro S (2001). Role of myosin light chain phosphorylation in the regulation of cytokinesis. *Cell Struct. Funct.*, **26**, 639–644.

Matsuzawa T, Fukui A, Kashimoto T, Nagao K, Oka K, Miyake M, and Horiguchi Y (2004). *Bordetella* dermonecrotic toxin undergoes proteolytic processing to be translocated from a dynamin-related endosome into the cytoplasma in an acidification-independent manner. *J. Biol. Chem.*, **279**, 2866–2872.

Matsuzawa, T, Kashimoto T, Katahira J, and Horiguchi Y (2002). Identification of a receptor-binding domain of *Bordetella* dermonecrotic toxin. *Infect. Immun.*, **70**, 3427–3432.

Michiels F, Habets G G M, Stam J C, Van der Kammen R A, and Collard J G (1995). A role for rac in Tiam1-induced membrane ruffling and invasion. *Nature*, **375**, 338–340.

Mills M, Meysick K C, and O'Brien A D (2000). Cytotoxic necrotizing factor type 1 of uropathogenic *Escherichia coli* kills cultured human uroepithelial 5637 cells by an apoptotic mechanism. *Infect. Immun.*, **68**, 5869–5880.

Minden A, Lin A, Claret F-X, Abo A, and Karin M (1995). Selective activation of the JNK signaling cascade and c-Jun transcriptional activity by the small GTPases Rac and Cdc42Hs. *Cell*, **81**, 1147–1157.

Nobes C D and Hall A (1999). Rho GTPases control polarity, protrusion, and adhesion during cell movement. *J. Cell Biol.*, **144**, 1235–1244.

Olson M F, Pasteris N G, Gorski J L, and Hall A (1996). Faciogenital dysplacia protein (FGD1) and Vav, two related proteins required for normal embryonic development, are upstream regulators of Rho GTPases. *Curr. Biol.*, **6**, 1628–1633.

Olson M F, Paterson H F, and Marshall C J (1998). Signals from Ras and Rho GTPases interact to regulate expression of p21[Wafl/Cip1]. *Nature*, **394**, 295–299.

Oshima M, Dinchuk J E, Kargman S L, Oshima H, Hancock B, Kwong E, Trzaskos J M, Evans J F, and Taketo M M (1996). Suppression of intestinal polyposis in Apc delta 716 knockout mice by inhibition of cyclooxygenase 2 (COX-2). *Cell*, **87**, 803–809.

Oswald E and de Rycke J (1990). A single protein of 110 kDa is associated with the multinucleating and necrotizing activity coded by the Vir plasmid of *Escherichia coli*. *FEMS Microbiol. Lett.*, **68**, 279–284.

Oswald E, de Rycke J, Guillot J F, and Boivin R (1989). Cytotoxic effect of multi-nucleation in HeLa cell cultures associated with the presence of Vir plasmid in *Escherichia coli* strains. *FEMS Microbiol. Lett.*, **58**, 95–100.

Oswald E, Sugai M, Labigne A, Wu H C, Fiorentini C, Boquet P, and O'Brien A D (1994). Cytotoxic necrotizing factor type 2 produced by virulent *Escherichia coli* modifies the small GTP-binding proteins Rho involved in assembly of actin stress fibers. *Proc. Natl. Acad. Sci. USA*, **91**, 3814–3818.

Page K, Li J, Hodge J A, Liu P T, Vanden Hoek T L, Becker L B, Pestell R G, Rosner M R, and Hershenson M B (1999). Characterization of a Rac1 signaling pathway to cyclin D_1 expression in airway smooth muscle cells. *J. Biol. Chem.*, **274**, 22065–22071.

Pedersen L C, Yee V C, Bishop P D, Trong I L, Teller D C, and Stenkamp R E (1994). Transglutaminase factor XIII uses proteinase-like catalytic triad to crosslink macromolecules. *Protein Sci.*, **3**, 1131–1135.

Pei S, Doye A, and Boquet P (2001). Mutation of specific acidic residues of the CNF1 T domain into lysine alters cell membrane translocation of the toxin. *Mol. Microbiol.*, **41**, 1237–1247.

Philips A, Roux P, Coulon V, Bellanger J-M, Vié A, Vignais M-L, and Blanchard J M (2000). Differential effect of Rac and Cdc42 on p38 kinase activity and cell cycle progression of nonadherent primary mouse fibroblasts. *J. Biol. Chem.*, **275**, 5911–5917.

Rippere-Lampe K E, Lang M, Ceri H, Olson M, Lockman H A, and O'Brien A D (2001a). Cytotoxic necrotizing factor type 1-positive *Escherichia coli* causes increased inflammation and tissue damage to the prostate in a rat prostatitis model. *Infect. Immun.*, **69**, 6515–6519.

Rippere-Lampe K E, O'Brien A D, Conran R, and Lockman H A (2001b). Mutation of the gene encoding cytotoxic necrotizing factor type 1 (cnf_1) attenuates the virulence of uropathogenic *Escherichia coli*. *Infect. Immun.*, **69**, 3954–3964.

Schmidt G, Goehring U-M, Schirmer J, Lerm M, and Aktories K (1999). Identification of the C-terminal part of *Bordetella* dermonecrotic toxin as a transglutaminase for Rho GTPases. *J. Biol. Chem.*, **274**, 31875–31881.

Schmidt G, Goehring U-M, Schirmer J, Uttenweiler-Joseph S, Wilm M, Lohmann M, Giese A, Schmalzing G, and Aktories K (2001). Lysine and polyamines are substrates for transglutamination of Rho by the *Bordetella* dermonecrotic toxin. *Infect. Immun.*, **69**, 7663–7670.

Schmidt G, Selzer J, Lerm M, and Aktories K (1998). The Rho-deamidating cytotoxic-necrotizing factor CNF1 from *Escherichia coli* possesses transglutaminase activity – cysteine-866 and histidine-881 are essential for enzyme activity. *J. Biol. Chem.*, **273**, 13669–13674.

Sehr P, Joseph G, Genth H, Just I, Pick E, and Aktories K (1998). Glucosylation and ADP-ribosylation of Rho proteins – Effects on nucleotide binding, GTPase activity, and effector-coupling. *Biochemistry US*, **37**, 5296–5304.

Sekine A, Fujiwara M, and Narumiya S (1989). Asparagine residue in the rho gene product is the modification site for botulinum ADP-ribosyltransferase. *J. Biol. Chem.*, **264**, 8602–8605.

Symons M and Settleman J (2000). Rho family GTPases: More than just simple switches. *Trends Cell Biol.*, **10**, 415–419.

Thomas W, Ascott Z K, Harmey D, Slice L W, Rozengurt E, and Lax A J (2001). Cytotoxic necrotizing factor from *Escherichia coli* induces RhoA-dependent expression of the cyclooxygenase-2 gene. *Infect. Immun.*, **69**, 6839–6845.

Walker K E and Weiss A A (1994). Characterization of the dermonecrotic toxin in members of the genus *Bordetella*. *Infect. Immun.*, **62**, 3817–3828.

Westwick J K, Lambert Q T, Clark G J, Symons M, Van Aelst L, Pestell R G, and Der C J (1997). Rac regulation of transformation, gene expression, and actin organization by multiple, PAK-independent pathways. *Mol. Cell. Biol.*, **17**, 1324–1335.

Wilde C and Aktories K (2001). The Rho-ADP-ribosylating C3 exoenzyme from *Clostridium botulinum* and related C3-like transferases. *Toxicon*, **39**, 1647–1660.

Wilde C, Chhatwal G S, and Aktories K (2001). C3stau, a new member of the family of C3-like ADP-ribosyltransferases. *Trends Microbiol.*, **10**, 5–7.

Würtele M, Renault L, Barbieri J T, Wittinghofer A, and Wolf E (2001). Structure of ExoS GTPase activating domain. *FEBS Lett.*, **491**, 26–29.

Zheng Y, Olson M F, Hall A, Cerione R A, and Toksoz D (1995). Direct involvement of the small GTP-binding protein Rho in *lbc* oncogene function. *J. Biol. Chem.*, **270**, 9031–9034.

CHAPTER 4

Cytolethal distending toxins: A paradigm for bacterial cyclostatins

Bernard Ducommun and Jean De Rycke

During the last 10 years, information has accumulated showing that pathogenic bacteria can produce various proteins able to block the eukaryotic cell cycle or delay its progression. These observations raise the attractive hypothesis that control of cell proliferation is a real strategy of pathogenicity, giving an evolutionary advantage to bacteria, and not simply a fortuitous effect observable in cell cultures, the interest of which would eventually be confined to cellular biologists or pharmacologists. The ultimate objective of this chapter is to analyse critically the pertinence of this candidate concept within the field of cellular microbiology (Cossart et al., 1996; Henderson et al., 1998) and to propose tentative criteria to define what we suggest calling bacterial cyclostatins.

We have chosen cytolethal distending toxin (CDT) as a prototype cyclostatin. From a probable common ancestor, CDT has spread through the bacterial world and it can be found in several Gram-negative pathogenic bacteria, constituting a family of toxins sharing common molecular and biological properties in spite of a large genetic dispersion (De Rycke and Oswald, 2001). The presence of a CDT homologue in various unrelated bacterial species is peculiar and suggests that CDT confers a strong selective advantage to producing bacteria, possibly helping adaptation to the host and ecological niche, or increasing pathogenicity. Another major interest in using CDT as a prototype cyclostatin is that a consistent picture of its mode of action on mammalian cells is now emerging, as a result of intensive research effort in recent years by several teams of investigators. The expertise acquired with CDT may aid in the study of other putative bacterial cyclostatins which have been described in the literature.

We will first briefly review our current knowledge of the molecular processes controlling cell cycle progression and of the checkpoint mechanisms

a

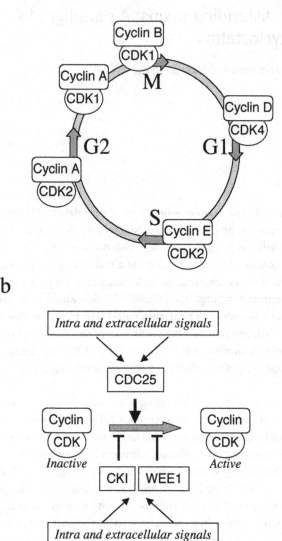

b

Figure 4.1. Cell Cycle control in animal cells. a) Cyclin-dependent kinases complexes. CDK/cyclins complexes regulate progression during the cell cycle (Nurse, 1990). Mitotic cyclins A and B accumulate throughout interphase, peaking at mitosis before being actively degraded by a mechanism that is dependent on the proteasome pathway. Other cyclins accumulate and peak during specific windows of the cell cycle; for example, cyclins D are expressed in G1 and degraded upon entry into S-phase. The CDK4/cyclins D as well as CDK2/cyclin E complexes phosphorylate substrates such as the pRB protein (Sherr,

that ensure the fidelity of the cell cycle and its adaptation to intra- and extra-cellular cues. We will then present the work from several laboratories that has been performed recently to examine the action of CDT and to identify its molecular targets. We will mainly focus on the biological and *in vivo* aspects of cell intoxication by CDT. At this point, we will compare CDT with other potential cyclostatins described in the literature. Finally, we will propose a general scheme to address the basic experimental cell biology questions that have to be answered in order to qualify a bacterial product as a true cyclostatin.

CELL CYCLE AND CHECKPOINTS

Proliferating eukaryotic cells divide into two daughter cells with strictly identical genomes. The central mechanism, at the heart of this process, is controlled by a family of enzymes, namely the Cyclin-Dependent Kinases or CDK, which is conserved in every eukaryotic organism (Pines, 1995). These kinases, in association with cyclins, regulate entry into mitosis and control the various steps of the progression through the cell cycle (Figure 4.1a). The activity of the CDK/cyclin complexes is tightly regulated and adapted to various intra- and extra-cellular parameters. First of all, the catalytic activity of CDK requires association with its cyclin. As their name implies, cyclins are proteins whose abundance varies during the cell cycle, being actively degraded by a mechanism that is dependent on the proteasome pathway. A balance between CDC25 phosphatases and WEE1 kinase activities is central to the accumulation of the CDK/cyclin complex in an inactive state, or in its activation and in the firing of its catalytic activity (Figure 4.1b). The regulation

---

1994), thus setting up the conditions required for progression into S-phase. CDK2/cyclin A is involved in the control of S-phase. CDK1/cyclin A and CDK1/cyclin B phosphorylation activities are essential for the major architectural and biochemical events leading to mitosis. b) Regulators of CDKs. The activity of CDC25 is essential to allow entry into mitosis because it activates the CDK1/Cyclin B kinase by dephosphorylation of two critical residues, threonine14 and tyrosine 15 (Nilsson and Hoffmann, 2000). These two positions are phosphorylated by WEE1 and related kinases that therefore play inhibitory roles on entry into mitosis (Rhind and Russell, 2001). In humans, the CDC25A phosphatase is involved in the activation of the CDK2/cyclin E complex, while the phosphatases CDC25B and CDC25C are responsible for the activation of the mitotic CDK/cyclin complexes. There are two families of Cyclin-dependent Kinases Inhibitors (CKI) in vertebrates. Members of the first one, including p21Cip1, p27Kip1 and p57Kip2, associate with the CDK/cyclin complex to inhibit its catalytic activity (Ball, 1997). The second family of inhibitors includes p15, p16, p18, p19 that act by competing with cyclin for association with CDK (Carnero and Hannon, 1998).

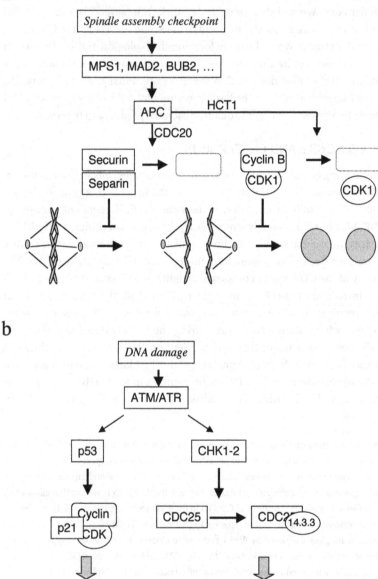

Figure 4.2. Checkpoints and targets. a) Mitotic checkpoint. Simplified view of the spindle assembly checkpoint. The activation of the mitotic checkpoint, also called Spindle Assembly Checkpoint (SAC), involves a signalling cascade implicating several actors that are partly identified (MPS1, MAD2, BUB2, . . .). Detection of spindle abnormalities or lack

of the activity of the CDK/cyclins complexes is also ensured by a family of inhibitory molecules called CKI, for Cyclin-dependent Kinase Inhibitors (Sherr and Roberts, 1999). For instance, p21, the expression of which is dependent on the p53 tumour suppressor, arrests cell cycle progression by inhibiting the activity of the CDK/cyclin complexes.

How does the cell control those events that are essential for ensuring the integrity of the genome and for cell survival, such as the completion of DNA replication, the absence of DNA damage, and the integrity of chromosomes and their correct segregation? The simplest response is to block cell cycle progression when the slightest abnormality is discovered and at the same time to activate repair mechanisms. When the deleterious risk has been dealt with the cell cycle can then restart. It is thus essential that abnormalities and damage be efficiently detected, and the information correctly and efficiently transduced to the cell cycle machinery (Zhou and Elledge, 2000).

Mitotic spindle defects or the failure of chromosome attachment to the spindle activate a checkpoint that leads to cell cycle arrest at the transition between metaphase and anaphase, when the division of the genome into identical sets must be performed. This mitotic spindle assembly checkpoint (SAC) involves sensing proteins that are able to detect any tension abnormalities between chromosomes and microtubules (see Figure 4.2a for more details on the molecular aspects). Recently, Gachet et al. (2001) have described a new mitotic checkpoint that monitors the integrity of the actin cytoskeleton and delays sister chromatid separation, spindle elongation, and cytokinesis until spindle poles have been properly oriented (Gachet et al.,

---

of chromosome attachment leads to the inhibition of the activity of the Anaphase Promoting Complex (APC/cyclosome) (Morgan, 1999). This enzymatic complex with ubiquitin ligase activity activates the degradation of the molecules responsible for the cohesion between sister chromatids (CDC20 dependent), as well as the degradation of the cyclins (HCT1 dependant) to turn off the activity of mitotic CDK/cyclin complexes.
b) DNA damage checkpoint. Simplified view of the DNA damage activated checkpoint. Detection of DNA damage leads to the activation of ATM and ATR kinases. Phosphorylation and stabilisation of p53 result in the accumulation of the p21Cip1 inhibitor that subsequently leads to the inhibition of the activity of the CDK/cyclin complexes at the G1/S transition (Zhou and Elledge, 2000). Upon phosphorylation of CHK1 and CHK2 kinases by ATM/ATR, CDC25 is phosphorylated. This modification leads to the association to proteins of the 14.3.3 family and to its cytoplasmic retention (Bulavin et al., 2002; O'Connell et al., 2000). This might also be responsible for a decrease of the CDC25 phosphatase activity. This results in the inability of CDC25 to activate the CDK/cyclin complexes and to the arrest of the cycle at the G2/M transition.

2001). Whether this mitotic checkpoint exists in higher eukaryotes has not yet been demonstrated. Similarly, it has been reported in budding yeast that actin network disorganisation is able to trigger a G2 delay (Harrison et al., 2001). However, alteration of the actin cytoskeleton does not always result in the activation of a checkpoint mechanism and can give rise to a binucleated cell (Robinson and Spudich, 2000), as with toxins that affect Rho (see Chapter 3).

Upon detection of DNA damage or incomplete DNA replication, a signalling cascade involving the PI3 kinases ATM (Ataxia Telangiectasia Mutated) and ATR (Ataxia Telangiectasia and rad3-related kinase) is activated. These kinases phosphorylate various substrates, including the p53 tumour suppressor and the checkpoint kinases CHK1 and CHK2 (Walworth, 2001). As schematically presented in Figure 4.2b, different pathways involving the p21 inhibitor and the CDC25 phosphatases will thus be activated to stop the cell cycle. As anticipated from this model, cells deficient for p53, as is the case in a large number of tumour-derived models, will block their cell cycle less efficiently in G1 and will stop mostly at the G2/M transition.

Signalling pathways involving the ERKs, their regulators, and their substrates ensure that coordination exists between extracellular signals and the cell cycle machinery. The major impact of the extracellular signals occurs in G1 when the availability of growth factors is sensed, decoded, and transduced to allow an adapted cellular response such as the transcription and the accumulation of type D cyclins. These cyclins are then able to form complexes with CDK that drive the cell into S-phase, unless their activity is repressed by association with CKI (Sherr and Roberts, 1999). Thus, extracellular parameters are taken into account by the cell and converted into proliferative or antiproliferative information. For instance, NGF induces the expression of CKIs such as p21, leading to the inactivation of G1 CDK/cyclin complexes and resulting in cell cycle arrest prior to the initiation of neuronal differentiation (Billon et al., 1996). In some cases, it seems that growth factor signalling pathways are also able to target G2 events. For instance, high levels of Epidermal Growth Factor (EGF) have been shown to transiently inhibit the transition from G2 to mitosis (Kinzel et al., 1990). The molecular nature of that effect is not fully elucidated, although it has been suggested that EGF prevents CDC25C activation (Barth et al., 1996). Recently, it has been shown that ErbB2, a receptor tyrosine kinase belonging to the EGF-receptor subfamily, is able to bind and specifically phosphorylate CDK1 on Tyr 15, thus delaying entry into mitosis. Breast cancer cells and tumours have been shown to overexpress ErbB2, and this is suspected to contribute to their resistance to Taxol-induced apoptosis (Tan et al., 2002).

Each of the mechanisms and each of the steps in the signalling pathways described represents a potential target that a pathogenic organism can use to block or perturb the host cell cycle and favour its own proliferation.

CDT AS A PROTOTYPIC CYCLOSTATIN

CDT and the Cell Cycle: From G2 Arrest to DNA Damage Checkpoint

Cytolethal Distending Toxin (CDT) was first described in 1987 as an activity contained in *Escherichia coli* bacterial supernatants that caused progressive cell enlargement and eventual cell death (Johnson and Lior, 1988). CDT was subsequently identified in a range of unrelated bacterial species including *Shigella dysenteriae* (Okuda et al., 1997), *Actinobacillus actinomycetemcomitans* (Sugai et al., 1998), and *Haemophilus ducreyi* (Cope et al., 1997). Initial studies on the mode of action of CDT led to the first observation that CDT-intoxicated HeLa cells were arrested in the G2-phase of the cell cycle (Peres et al., 1997). The effect of CDT on the cell cycle of a large number of human cell lines has been investigated over the last few years. As summarised by Cortes-Bratti et al. (2001a), with the exception of human foreskin and embryonic lung fibroblasts, which are arrested in either G1- or G2-phases, all cell lines tested so far (including epithelial, B and T cells) arrested their cycle in G2 and eventually entered an apoptotic process (Cortes-Bratti et al., 2001b; De Rycke et al., 2000).

As the CDK1/cyclin B complex activity is a key player in the regulation of the G2-phase to mitosis transition, its activity in CDT-treated cells was examined. CDK1/cyclin B was found to be present in an inactive, hyper-phosphorylated state (Comayras et al., 1997). The CDK1/cyclin B complex retrieved from CDT-treated cells could be reactivated *in vitro* using recombinant CDC25 phosphatase (Sert et al., 1999), thus eliminating the possibility that association with a CKI such as p21Cip1 was responsible for CDK1 inactivation. *In vivo*, the accumulation of inactive CDK/cyclin B correlated with the exclusion of CDC25C from the nucleus (Alby et al., 2001). Expression of either CDC25B or CDC25C in CDT-treated HeLa cells reversed the cell cycle arrest, driving the cell into an abnormal mitotic process (Escalas et al., 2000). Taken together, these results strongly indicate that the CDT-induced cell cycle arrest was dependent on a regulatory event upstream of CDC25.

This phenotype is similar to that seen in G2 arrest activated by DNA damage – for example, in the ability to be rescued by caffeine (Sert et al., 1999). However, because initial studies using the "comet" assay were unable to detect DNA damage in CDT-treated cells (Sert et al., 1999), it was initially hypothesised that CDT was able to highjack this pathway and to illegitimately activate one of its steps. As already presented (Figure 4.2b), cell cycle arrest in G2 in response to DNA injury is dependent on the activation of a transduction cascade that includes ATM/ATR and the checkpoint 1/2 kinases. Studies performed in the authors' laboratories demonstrated that Chk2 kinase was indeed activated upon CDT intoxication in HeLa cells (Alby et al., 2001). The involvement of its upstream regulator, the ATM kinase, was suggested from work performed using ATM-deficient cells (Cortes-Bratti et al., 2001b; Li et al., 2002). In human fibroblasts that arrest their cell cycle in G1 in response to CDT treatment, p53 and its transcriptional target p21Cip1 were activated similarly to that observed upon treatment with ionising radiation (Cortes-Bratti et al., 2001b). Finally, it has recently been shown in HeLa cells that CDT induces phosphorylation of histone H2AX and relocalisation of the DNA repair complex Mre11, similarly to that observed after exposure to ionising radiation (Cortes-Bratti et al., 2001b; Li et al., 2002). The likely conclusion emerging from these studies is that CDT is indeed a DNA-damaging agent. A summary of the major findings depicting the effects of CDT on cell cycle control is shown in Figure 4.3.

CDT-B is the Catalytic Subunit

Of the three subunits CDT-A, CDT-B, and CDT-C making up the holotoxin, it is now established that CDT-B bears the catalytic activity (Elwell et al., 2001; Elwell and Dreyfus, 2000; Lara-Tejero and Galan, 2000). Cytosolic expression of CDT-B alone does reproduce the cytostatic effect of the holotoxin (Elwell et al., 2001; Lara-Tejero and Galan, 2000; Mao and DiRienzo, 2002). Although CDT subunits bear no significant sequence similarity with proteins present in databases, the putative nature of the CDT-B enzymatic property was approached using three-dimensional structure sequence analysis (De Rycke and Oswald, 2001; Elwell and Dreyfus, 2000; Lara-Tejero and Galan, 2000). The CDT-B structure has the best compatibility score with a broad family of enzymes sharing phosphodiesterase activity, such as human DNase-I, human DNA repair endonuclease Hap1, exonuclease III from *E. coli*, and also with certain sphingomyelinases and inositol phosphatases. Despite the lack of overall sequence identity with these proteins (less than 15%), most of the catalytic and ion metal binding sites are conserved. Moreover, mutation of these potentially critical residues abrogates the cytostatic activity of the toxin, which demonstrates unambiguously the role of the phosphodiesterase catalytic site in the cell cycle activity of the toxin (Elwell and Dreyfus, 2000; Lara-Tejero and Galan, 2000).

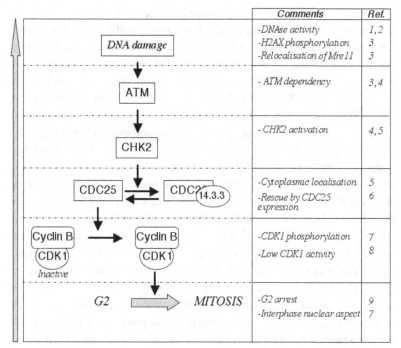

(1) Elwell and Dreyfus, 2000; (2) Lara-Tejero and Galan, 2000; (3) Li et al. 2002; (4) Cortes-Bratti et al., 2001b; (5) Alby et al., 2001; (6) Escalas et al., 2000; (7) Comayras et al., 1997; (8) Sert et al., 1999, (9) Peres et al., 1997.

Figure 4.3. CDT and cell cycle control. Involvement of the actors of the DNA damage checkpoint cascade was investigated, from the first observation (bottom) to the recent evidence for a DNA damaging activity (top).

Based on the nature of the conserved catalytic residues, CDT-B is not more closely related to DNase-I than to the other phosphodiesterases mentioned above (De Rycke and Oswald, 2001). Moreover, as developed later, DNA damage is not the only primary event able to trigger a signal cascade leading to a cell cycle block in G2. However, several complementary observations suggest that nuclear DNA is the primary target of CDT-B. (1) Firstly, nuclear translocation was demonstrated in COS-1 cells either transfected with a plasmid encoding tagged CDT-B, or after microinjection of the purified toxin (Lara-Tejero and Galan, 2000). In the above experiments, CDT-B transient expression caused marked chromatin disruption, which was abrogated with CDT-B mutants that had substitutions in residues putatively required for catalysis or magnesium binding. (2) Secondly, this observation was confirmed in a yeast model, where ectopic expression of CDT-B recapitulates the major effects observed with CDT-treated mammalian cells together with an extensive chromosome degradation, occurring as early as 4 h after CDT-B expression (Hassane et al., 2001). Here again, the effects of CDT-B were dependent upon the integrity of the putative catalytic sites of the molecule. (3) Thirdly, purified CDT-B or holotoxin used at very high concentration was also reported to cause a DNA-nicking effect on supercoiled plasmid DNA *in vitro* (Elwell et al., 2001). (4) Lastly, highly concentrated *Haemophilus ducreyi* CDT was shown to induce double-strand breaks in culture cells after 8 hours of exposure, as detected in pulsed field gel electrophoresis (Frisan et al., 2003).

Although the above results are consistent with the hypothesis that DNA is the primary target of CDT-B, caution must be exercised as to the actual relevance to "physiological" exposure. All the results described above were obtained in somewhat extreme conditions, i.e., with a toxin concentration probably far above that required to trigger the cell cycle block. In contrast, no detectable genomic alteration has been observed in mammalian cells exposed to closer-to-physiological doses causing total cell cycle block in G2 (Sert et al., 1999). These apparent discrepancies between concentrations sufficient to induce the G2 block and those required to cause genomic alteration warrant closer attention in future experiments, with a view to establishing a firmer basis for the causal relationship between the two processes. Further, if DNA is really a natural target for CDT, another critical issue to clarify is the binding of the toxin to the DNA, as observed with other nucleases but not yet reported for CDT at the time of submission.

If CDT-B is considered as the catalytic subunit of the holotoxin that accounts for the specific effect on the cell cycle, it should be emphasised that some investigators have also attributed cytotoxic or cytostatic activity to *A. actinomycetemcomitans* recombinant CDT-C. A cell-blocking effect was noted in PHA-activated human T cells following external exposure (Shenker et al., 2000), while cytotoxicity was observed in Chinese hamster ovary (CHO) after cytosol delivery (Mao and DiRienzo, 2002). These results disagree with other studies showing that, unlike CDT-B, internal expression of *C. jejuni* CDT-C in mammalian (Lara-Tejero and Galan, 2000) or yeast cells (Hassane et al., 2001) does not induce significant cytotoxic or cytostatic effects. Further investigation is therefore needed to clarify the possible contribution of CDT-C to the cell cycle effect of the holotoxin. Such studies should define the exact modality of this cell toxicity and whether it is dependent or not on a specific catalytic activity (as is the case for CDT-B) or is the result of an indirect toxic effect observed with a high concentration of the protein.

Aside from the catalytic activity of the toxin, which resides on CDT-B, a specific function has not yet been clearly assigned to CDT-A or CDT-C, whose presence is generally deemed essential in the case of external exposure. Their indispensable role together with CDT-B is demonstrated by genetic evidence, as the three *cdt* genes are required concomitantly to determine the toxic phenotype (Peres et al., 1997; Pickett et al., 1994; Scott and Kaper, 1994; Sugai et al., 1998) and by reconstitution of CDT activity using individually purified CDT subunits (Deng et al., 2001; Frisk et al., 2001; Lara-Tejero and Galan, 2001; Lewis et al., 2001; Saiki et al., 2001). From these results, it is generally postulated that the three CDT proteins form a tripartite complex required for toxicity, a conclusion strongly supported by gel filtration chromatography (Lara-Tejero and Galan, 2001).

Because CDT-B, once internalised in target cells, is able to reproduce all the effects of the holotoxin, it is likely that neither CDT-A nor CDT-C is required during the late stages of cellular trafficking, in particular for nuclear translocation and binding to a nuclear target. The general model of AB toxins, where A refers to the active subunit and B to the subunit(s) mediating binding to receptors and translocation across the cell membrane, can therefore be tentatively applied to CDT. According to such a model, CDT-A and CDT-C would fit the B module required for the delivery of CDT-B, the A module. A CDT-A contribution to binding of the holotoxin is suggested by the existence of an AA motif similar to a lectin fold present in the B chain of two AB toxins from plants: abrin and ricin (Lara-Tejero and Galan, 2001). Furthermore, the recombinant product of the *cdtA* gene of *A. actinomycetemcomitans* binds significantly to the surface of sensitive mammalian cells, whereas the products of *cdtB* and *cdtC* do not (Mao and DiRienzo, 2002).

The AB model does not fit an observation with human T cells. Cell extracts containing recombinant CDT-B from *A. actinomycetemcomitans* alone (called ISF for immunosuppressive factor) have the capacity to induce a G2 arrest in PHA-activated human T cells upon external exposure (Shenker et al., 2000). To account for this exception, one possibility is that activated lymphocytes have an innate capacity to internalise CDT-B, thus bypassing the requirement of the delivery stage by CDT-A and CDT-C.

In Vivo Relevance of the Anti-Proliferative Activity of *CDT*

Current knowledge about the mode of action of CDT comes mainly from studies in cell cultures. The exact contribution of CDT to the pathogenicity of the producing organisms is still therefore largely speculative. A major challenge now is to examine to what extent the concept of a cyclostatin applies *in vivo*, in other words (1) whether CDT really does contribute to the control of proliferation of specific cell populations in infected hosts and (2) what the pathogenic consequences are of such control.

A preliminary question is whether the putative genotoxicity of CDT can be viewed as a potential mechanism of pathogenicity *per se*. In our opinion, this is unlikely, as the DNA damage caused by CDT under physiological cell exposure is mild or undetectable and does not induce signs of early lethality in cell cultures (Lara-Tejero and Galan, 2002; Sert et al., 1999). The delayed cytotoxic effects are clearly a consequence of the cell cycle block, whether it is in epithelial cells (Cortes-Bratti et al., 2001a; Cortes-Bratti et al., 2001b; De Rycke et al., 2000) or in lymphocytes where apoptosis is observed (Cortes-Bratti et al., 2001b; Shenker et al., 2001). Furthermore, CDT cytotoxicity appears essentially restricted to proliferating cells because no effect is detectable in confluent epithelial or fibroblastic cell lines (Johnson and Lior [1988] and personal observations), and because cells must transit through an S-phase to be committed to the G2 arrest. However, as suggested by Li et al. (2002), non-proliferating dendritic cells can also be targeted by CDT to induce DNA damage.

We can therefore speculate that the evolutionary advantage conferred by CDT on producing bacteria is related to the conferred ability to modulate the growth of infected tissues. The consequences of CDT activity should be predictably much more significant in tissues where the process of cell proliferation is essential such as epithelial growth, cell immune response, or wound healing. Mucous barriers, in particular intestinal epithelia, which are characterised by a rapid renewal of enterocytes from crypt cells and also by the presence of enormous numbers of intraepithelial lymphocytes, are candidate target tissues for CDT. This prediction is consistent with the observation that CDT-producing bacterial species are known to efficiently colonise at various mucous barriers: digestive tract for *E. coli, Campylobacter sp., Helicobacter sp.*, and *Shigella sp.*, periodontal pocket for *A. actinomycetemcomitans*, and genital region for *H. ducreyi*.

Studies on primary cells constitute a further step toward the understanding of CDT impact *in vivo*. In particular, the effect of CDT on human peripheral blood mononuclear cells, including B and T lymphocytes, provides a relevant model of the impaired local immune response. Published results clearly show an effective inhibition of the proliferative immune response following stimulation by various mitogens, particularly with CDT from *A. actinomycetemcomitans* (Shenker et al., 2000; Shenker et al., 1999) and from *H. ducreyi* (Gelfanova et al., 1999; Svensson et al., 2001). The potential impact of CDT on the immune response is highly relevant in patients infected with either of these two agents. In the case of severe impairment of the local immune response, localised periodontal disease due to *A. actinomycetemcomitans* can more easily develop into a generalised infection including endocarditis, meningitis, and osteomyelitis (Wilson and Henderson, 1995). In the same way, a major component of the immune response to

H. ducreyi is a T-cell infiltration at the site of infection, which is maintained throughout the pustular then ulcerative stages of the disease. Inhibition of local T-cell growth by CDT could help *H. ducreyi* evade the immune response and propagate outside the initial lesion, or could facilitate recurrence of the disease (Gelfanova et al., 1999; King et al., 1996).

To our knowledge, the direct impact of CDT on cell proliferation in animals experimentally exposed to CDT or to CDT-producing organisms has never been reported. Published observations in enteric mouse models of infection give some insights into the possible contribution of CDT to pathogenesis but do not directly address its effect on the proliferation of enterocytes or intraepithelial lymphocytes. In a suckling mouse model, CDT from *S. dysenteriae* administrated orally was shown to induce diarrhoea (Okuda et al., 1997). The microscopic lesions reported in the descending colon shortly after administration consisted of necrosis and reparative hyperplasia, neither of which evokes an antiproliferative effect. In another study, the effects of oral administration of a *C. jejuni* strain and of isogenic *cdtB* mutants to immunodeficient mice were compared. Mutant strains were unaffected in enteric colonisation but partially lost their ability to translocate into blood, spleen, and liver (Purdy et al., 2000). This study suggests that CDT may contribute to the invasiveness of the challenge strain, but the relationship between this property and a possible blocking or differentiating effect on enterocytes was not addressed. Finally, the role of CDT in the formation of chancroid ulcer, a genital lesion caused by *H. ducreyi*, has been investigated using two experimental skin models of infection in humans and in rabbits. These models reproduce the early stages of the disease, up to the formation of pustules (Lewis et al., 2001; Stevens et al., 1999; Young et al., 2001). CDT was clearly not required for the formation of pustules, but its participation in the formation of ulcers and in the retardation of healing, both of which characterise the later stages of the disease, remains open to question.

EXAMPLES OF POTENTIAL MEMBERS OF THE CYCLOSTATIN FAMILY

We have limited our field of interest to bacteria, although other pathogens, such as viruses, fungi, and protozoa, may also express proteins controlling the eukaryotic cell cycle (Henderson et al., 1998; Op De Beeck and Caillet-Fauquet, 1997). CDT is but one among various bacterial protein products reported to exert an antiproliferative effect in cell cultures, as shown in Table 4.1. As an objective of this chapter is to propose a specific definition

Table 4.1 *Putative members of the cyclostatin family*[1,2]

Protein	Organism	Cells tested	Cell cycle effect	Cycle effectors	Postulated molecular target	Reference
CDT	Several Gram negative	Numerous epithelial and fibroblastic cell lines	G2 block	CDK1	DNA alteration	(De Rycke and Oswald, 2001)
		Activated HPBMC[2]/T and B lymphocytes	G2 block + apoptosis			
C2	*Clostridium botulinum*	HeLa epithelial cells	G2 block	CDK1	Activation of PP2A[2] Cytoskeleton checkpoint	(Barth et al., 1999)
CIF	*E. coli, Citrobacter rodentium*	HeLa epithelial cells	G2 block	CDK1	?	(Marches et al., 2003; Nougayrede et al., 1999)
FIP	*Fusobacterium nucleatum*	Activated HPBMC/T and B lymphocytes	Early/Mid G1	cyc D2/cyc D3	?	(Demuth et al., 1996; Shenker and Datar, 1995)
CNF	*E. coli, Yersinia pseudotuberculosis*	Epithelial, fibroblastic and megacariocytic cell lines	Inhibition of cytokinesis Polyploïdy	?	Cytoskeleton checkpoint ?Activation of Rho GTPase	(De Rycke et al., 1996; Denko et al., 1997; Hudson et al., 1996)
SAGP	*Streptococcus pyogenes*	Human and mice carcinoma	G2 block	CDK1	Dephosphorylation EGFR[2] Inhibition op42/44 MAPK[2]	(Yoshida et al., 2001)

VT	*E. coli*	Human mesangial cells	S (growth retardation)	?	Signal from Gb3[2] receptor ?	(Brigotti et al., 2002; van Setten et al., 1997)
ST	*E. coli*	Human colon carcinoma	S (growth retardation)	?	Activation of membrane guanylyl cyclase ?	(Pitari et al., 2001)
LIF	*E. coli*	Activated HPBMC[2]/T and B lymphocytes (Epithelial cells are not susceptible)	?	?	?	(Klapproth et al., 2000)
STI	*Salmonella typhimurium*	Mouse T cells unresponsive to Il-2	?	?	?	(Matsui et al., 1998)
PIP	*Helicobacter pylori*	Various epithelial and immune cells line	?	?	?	(Knipp et al., 1996)
.....	*Prevotella intermedia*	Activated T and B lymphocytes	?	?	?	(Shenker et al., 1991)

[1] *Acronyms for the bacterial proteins*: CDT: Cytolethal Distending Toxin; CIF: Cycle Inhibiting Factor; FIP: Fusobacterium Immunosuppressive Protein; CNF: Cytotoxic Necrotizing Factor; SAGP: Streptococcal Antitumor GlycoProtein; VT: Verotoxin; SLT: Shiga-Like Toxin; ST: thermostable enterotoxin; LIF: Lymphocyte Inhibiting Factor or lymphostatin; STI: *Salmonella typhimurium*-derived T-cell inhibitor; PIP: Proliferation Inhibiting Protein.

[2] *Other acronyms*: HPBMC: human peripheral blood mononuclear cells; PP2A: protein phosphatase 2A; EGFR: epidermal growth factor receptor; MAPK: mitogen-activated protein kinase; Gb3: globotriaosyl ceramide.

of cyclostatins, we have sought to highlight products strongly suspected of targeting signal transduction pathways that are directly linked to cell cycle control. Given the extreme refinement of cell cycle regulation, it is indeed likely that a number of other bacterial factors causing cellular stresses or injuries are able to induce non-specific delay of cellular growth. The selected factors are displayed according to their phenotypic effect on the cell cycle and, insofar as information is available, to their speculated mode of action at the molecular level.

Like CDTs, two other factors have been reported to cause a cell cycle block in G2: C2 toxin from *Clostridium botulinum* (Barth et al., 1999), and CIF (cycle inhibiting factor) from enteropathogenic *E. coli* and *Citrobacter rodentium* (Marches et al., 2003; Nougayrede et al., 1999). Although the final steps of the G2 checkpoint activated by these products are identical, namely the inactivation of CDK1 by tyrosine phosphorylation (Figure 4.1b), the upstream signalling cascades leading to CDK1 are probably different. The basic effect of C2 toxin – its ability to ADP-ribosylate monomeric actin – results in a total breakdown of the cell cytoskeleton. The activities of both CDK1 and CDC25-C phosphatase were reported to decrease after exposure to C2, but upstream events have not yet been investigated. Two original hypotheses have been raised: (1) direct ability of C2 to activate protein tyrosine phosphatases (Prepens et al., 1998) leading to the activation of protein phosphatase 2A (PP2A), resulting in the observed inactivation of CDC25-C; (2) activation of a postulated G2 mammalian actin cytoskeleton checkpoint, such as described in yeast (Harrison et al., 2001). However, a DNA damage-related G2 checkpoint cannot be dismissed even though C2 did not cause delayed progression in S-phase. CIF is a type III–secretion dependent protein (see Chapter 6) that also alters the actin cytoskeleton but, in contrast to C2, functions by inducing a permanent activation of stress fibres and focal adhesions (Marches et al., 2003; Nougayrede et al., 1999). Neither the events upstream of CDK1 nor the primary target and mode of action of CIF are known.

E. coli CNFs cause an irreversible induction of actin stress fibers through the constitutive activation of the small GTPases of the Rho family (reviewed by Boquet [2001] and Horiguchi [2001], and in Chapter 3). The causal relationship between Rho-GTPases induction and the permanent cytoskeleton reorganisation is straightforward, but the link between these events and the cell cycle block has not yet been investigated. CNF abrogates cytoplasmic division (or cytokinesis) without interfering with normal nuclear cycling, such that DNA is replicated in the absence of cell division (Denko et al., 1997). Phenotypically, blocked cells are thus both polyploïd (DNA content up to 16n) and multinucleated (De Rycke et al., 1996). A tentative hypothesis is that

CNF directly interferes with a Rho-GTPase that is required for contractile ring assembly at cell division (Ridley, 1995).

FIP (*F. nucleatum* immunosuppressive protein) from *Fusobacterium nucleatum* produces a G0/G1 block in activated lymphocytes (Demuth et al., 1996; Shenker and Datar, 1995). This activity has not been reported in epithelial or fibroblastic cells. A detailed analysis of cyclin expression showed that FIP does not prevent entry into G1, and that the cell cycle event sensitive to FIP resides somewhere between the restriction point of cyclin D2 (early to mid-G1) and that of cyclins D3 and E (mid- to late G1) (Shenker and Datar, 1995).

SAGP (streptococcal acidic glycoprotein) is an acidic glycoprotein from the *Streptococcus pyogenes* Su strain, originally identified as having anti-tumour activity in mice and being able to stop in a reversible manner the proliferation of several human and mice cell lines, particularly carcinoma cells (Yoshida et al., 1998). In A431 epithelioid cells, the growth-inhibitory effect of SAGP is associated with tyrosine dephosphorylation of the epidermal growth factor receptor (EGFR) followed by the inhibition of the p42/44 mitogen-activated protein kinase (MAPK) (Yoshida et al., 2001). It would be interesting to know the precise effect of SAGP on cell cycle parameters, in particular if the cells are delayed in G2, as observed for instance in A431 cells exposed to the epidermal growth factor itself (Kaszkin et al., 1996). In any case, these findings indicate that some bacteria are able to control proliferative signalling through G proteins and MAPK signalling (Gutkind, 1998).

STa (stable toxin a) and VT (verotoxin) are two famous prototypes *E. coli* toxins not originally described as growth inhibitory products. Recent studies show, however, that their enzymatic or binding properties may account for a major antiproliferative effect in cell lines and experimental settings intended to mimic the *in vivo* situation. VT is the main factor of pathogenicity of *E. coli* strains, causing the deadly haemolytic and uraemic syndrome (HUS). It is generally described as cytotoxic – able to rapidly kill standard cell cultures through inhibition of protein synthesis. In primary cultures of glomerular mesangial cells, which constitute its primary targets *in vivo*, VT does not affect cell viability but markedly inhibits DNA synthesis and proliferation and, to a much lesser extent, protein synthesis. The binding subunit is reported to contribute significantly to the growth inhibitory effect, hypothetically through a pathway originating from the glycosphingolipidic receptor globotriaosylceramide Gb3 (Van Setten et al., 1997). STa is the major determinant of enterotoxigenic *E. coli* causing travellers' diarrhoea. STa, which acts as an agonist of guanylyl cyclase (GC-C), its cell membrane receptor in enterocytes, has been neither described as a cytotoxin nor detected in the laboratory using such a property. Recently it was shown, however, that STa was able to strongly regulate cell cycle progression of human colon carcinoma cells through cyclic GMP-dependent mechanisms (Pitari et al., 2001).

Four further diversely characterised factors have been mentioned, namely *E. coli* LIF (leukaemia inhibitory factor), STI (*Salmonella typhimurium*-derived T-cell inhibitor), *Helicobacter pylori* PIP (proliferation-inhibiting protein), and a *Prevotella intermedia* protein (Klapproth et al., 2000; Knipp et al., 1996; Matsui et al., 1998; Shenker et al., 1991). Although their specific effect on the cell cycle has not been reported, they have in common the ability to inhibit the proliferation of immune cells.

HOW TO IDENTIFY THE TARGET OF A CYCLOSTATIN?

Do these bacterial products target cell cycle control? Do they affect the function, the expression, or the activity of cell cycle regulators? Do they reveal new regulatory pathways that remain to be uncovered? As reported above with CDT, the starting point of the investigation to address these questions is usually the result of two experiments. One is the examination of nuclear aspects by fluorescence microscopy to determine whether the cells are blocked or not in mitosis (i.e., condensed chromatin, metaphase plate, etc.). The second is the identification of a cell cycle arrest by flow cytometric analysis of DNA content. When these questions have been answered, we propose that a step-by-step investigation strategy should be followed as outlined in Figure 4.4. The first step (Figure 4.4a) leads to four different possibilities; the cells might be stacked in mitosis with a G2 DNA content and condensed chromosomes, or in interphase with a DNA content that reveals a G1, S, or G2 cell cycle arrest.

Arrest of the cell cycle with a G1 DNA content indicates that mechanisms involved in G1 progression or in the control of the G1/S transition are targeted. As presented in Figure 4.4b, monitoring the kinase activity associated with the CDK4/cyclin D complexes and with the CDK2/cyclin E complex is the first step. Inactivation of these two types of complexes might reflect an absence of association with their cyclin regulatory subunit, perhaps because its expression is inhibited. Interaction with members of the cyclin-dependent kinase inhibitors (CKI) family, such as p16 or p27, might also be responsible for inhibition of the kinase. In the absence of active CDC25A phosphatase, the CDK2 and CDK4 kinase will remain phosphorylated on tyrosine and catalytically inactive. Lack of expression or targeted degradation of CDC25A, which has been demonstrated following DNA damage (Mailand et al., 2000), might therefore also be involved. The *F. nucleatum* FIP toxin presented in

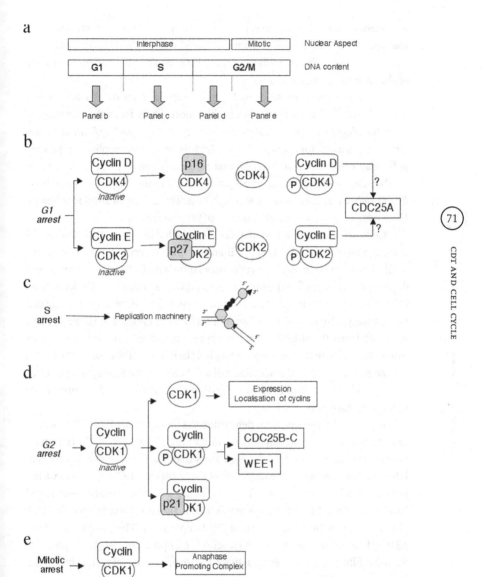

Figure 4.4. How to identify the target of Cyclostatin? Panel a depicts the basic steps in the analysis of the effects of a newly identified bacterial product that impairs proliferation and is suspected to interfere with the cell cycle machinery. Panels b–e illustrate the successive steps and the basic questions that have to be successively addressed to validate the bacterial product as a member of the cyclostatin family and to identify its molecular target.

the previous section leads to cell cycle arrest in G1. However, critical information such as the kinase activity associated with cyclin D2 complexes and the association with the inhibitor p27Kip1are required to confirm the nature of the molecular target.

Cell cycle arrest or delay in S phase (Figure 4.4c) may result from a direct effect on cell cycle regulators, but more likely from the targeting of essential players in the replication machinery, such as DNA polymerases and auxiliary proteins, for example PCNA. Impairment of the replication process will delay progression in S-phase and ultimately result in the activation of a DNA damage–dependent checkpoint. The VT and the STa toxins from *E. coli* may exert their actions through targeting the replication machinery. However, their molecular effects are not yet elucidated.

Cells might also be arrested in G2. As depicted in Figure 4.4d, this results from the absence of CDK1/cyclin B and/or CDK1/cyclin A activities, which might be due to (1) the lack of cyclin association with CDK1, (2) the absence of dephosphorylation of CDK1 on Tyr 15, or (3) the association of CDK1 with an inhibitor such as p21Cip1. These various hypotheses have to be investigated successively. The lack of association of cyclin A or cyclin B to CDK1 is likely to result from the inhibition of its expression or to a modification of its intracellular localisation. Hyperphosphorylation of CDK1 on Tyr 15 is of course reminiscent of the activation of a G2 checkpoint in response to DNA damage, which will lead to an investigation of the activity and features of the upstream regulatory elements.

Immunofluorescence can determine the localisation of CDC25C, whose cytoplasmic retention is dependent on its phosphorylation by CHK1 and 2 checkpoint kinases. Each of the regulators that play a role in the detection of DNA damage and in the transduction of the signal to CDC25 is possibly a target for a hijacking mechanism. The DNA damage cascade can also be activated from its origin, through the generation of a signal related to specific DNA alteration, or to the perturbation of DNA replication. The association of the p21Cip1 inhibitor with the CDK1/cyclin B complex can also be responsible for the inhibition of its catalytic activity, resulting in cell cycle arrest in late G2. Induction of p21Cip1 is dependent on either p53 dependent or independent pathways and has been shown to accumulate in the cell following the activation of various signalling pathways such as DNA damage or growth factor cascade. Thus, if the presence of p21Cip1 in association with the CDK1/cyclin B complex is revealed, it will be necessary to carry out a study of its expression to identify the target of the toxin. For instance, in the case of the CIF toxin, it has been shown that cyclin expression is not affected (Marches et al., 2003), leaving open the possibility that the CDK/cyclin complexes might be inactivated either by a CKI binding or by tyrosine phosphorylation. To date, there is no example for a direct action of a bacterial cyclostatin on CDC25 phosphatases or CHK kinases. However, it has been proposed that the C2 toxin activates PP2A, thus dephosphorylating and inactivating CDC25C.

As discussed above, cell cycle arrest in mitosis is a typical feature of the activation of the metaphase/anaphase checkpoint. Any alteration of the mitotic spindle assembly or misalignment of chromosomes will potentially be responsible for this activation. Upstream regulators of the mitotic checkpoint are being progressively identified (Figure 4.2b). One can therefore hypothesize that a critical step in this signal pathway, such as MAD2 protein or the MPS1 kinase, might also be targeted.

OUTLOOK: THE CHALLENGING CONCEPT OF CYCLOSTATINS

The overview of bacterial proteins blocking the cell cycle hints at the probably large diversity of their modes of action on the cell cycle, and highlights the scarcity of data on the mechanisms of action themselves, except for CDT. Moreover, the potential antiproliferative effect of these proteins has not been investigated *in vivo*, particularly in relation to the pathogenicity of producing strains, which limits any speculation about the biological outcome from such products and particularly about the selective advantage that they may confer to bacteria. It is therefore premature, in the present stage of knowledge, to propose a unifying view of the cyclostatin concept, which remains a working hypothesis based on the identification of intrinsic biological activity and, as such, needs to be challenged, particularly *in vivo*, using proper experimental designs.

Keeping in mind the above limitations, it is possible to propose preliminary criteria to define cyclostatins and classify them according to their specific effects on the cell cycle in cell cultures. The simplest and least restrictive definition would be to include in the cyclostatin family all the bacterial products that lead the intoxicated cells to stop their cell cycle at a specific point. Indeed, the concept of an identical execution point was central to the identification of the cell division cycle (cdc) regulatory genes (Hartwell, 1978). In a more precise definition, members of the cyclostatins would be only those bacterial products that specifically target a key regulator of the cell cycle machinery. To be less restrictive and in order to include toxins such as CDT, we propose to rephrase and slightly enlarge this definition by stating that a bacterial cyclostatin is a bacterial product that directly targets or interferes with one of the pathways that control the activity of cell cycle regulators in response to intra- and extra-cellular signals.

Besides a rigorous analysis of the cell cycle effects of candidate cyclostatins in cell cultures as stressed above, the hypothetical antiproliferative effect of these products should be challenged in appropriate animal models, either through inoculation of cell-free products, or, ideally, following infection with the producing bacteria. This is a primary condition for cyclostatins to emerge as a relevant subject in the field of cellular microbiology. The *in vivo* validation is all the more important because, in contrast to the classical concepts of cellular microbiology, the original observations were not reported in diseased natural hosts, but in cell cultures. This essential requirement applies in particular to CDT whose unique spreading among unrelated pathogenic bacterial species is a strong argument for a major role as a factor of pathogenicity or of adaptation to the host.

REFERENCES

Alby F, Mazars R, de Rycke J, Guillou E, Baldin V, Darbon J M, and Ducommun B (2001). Study of the cytolethal distending toxin (CDT)-activated cell cycle checkpoint. Involvement of the CHK2 kinase. *FEBS Lett.*, **491**, 261–265.

Ball K L (1997). p21: Structure and functions associated with cyclin-CDK binding. *Prog. Cell Cycle Res.*, **3**, 125–134.

Barth H, Hoffmann I, Klein S, Kaszkin M, Richards J, and Kinzel V (1996). Role of cdc25-C phosphatase in the immediate G2 delay induced by the exogenous factors epidermal growth factor and phorbolester. *J. Cell Physiol.*, **168**, 589–599.

Barth H, Klingler M, Aktories K, and Kinzel V (1999). *Clostridium botulinum* C2 toxin delays entry into mitosis and activation of p34cdc2 kinase and cdc25-C phosphatase in HeLa cells. *Infect. Immun.*, **67**, 5083–5090.

Billon N, van Grunsven L A, and Rudkin B B (1996). The CDK inhibitor p21WAF1/Cip1 is induced through a p300-dependent mechanism during NGF-mediated neuronal differentiation of PC12 cells. *Oncogene*, **13**, 2047–2054.

Boquet P (2001). The cytotoxic necrotizing factor 1 (CNF1) from *Escherichia coli*. *Toxicon*, **39**, 1673–1680.

Brigotti M, Alfieri R, Sestili P, Bonelli M, Petronini P G, Guidarelli A, Barbieri L, Stirpe F, and Sperti S (2002). Damage to nuclear DNA induced by Shiga toxin 1 and ricin in human endothelial cells. *Faseb J.* **16**, 365–372.

Bulavin D V, Amundson S A, and Fornace A J (2002). p38 and Chk1 kinases: Different conductors for the G(2)/M checkpoint symphony. *Curr. Opin. Genet. Dev.*, **12**, 92–97.

Carnero A and Hannon G J (1998). The INK4 family of CDK inhibitors. *Curr. Top. Microbiol.*, **227**, 43–55.

Comayras C, Tasca C, Peres S Y, Ducommun B, Oswald E, and De Rycke J (1997). *Escherichia coli* cytolethal distending toxin blocks the HeLa cell cycle at the G2/M transition by preventing cdc2 protein kinase dephosphorylation and activation. *Infect. Immun.*, **65**, 5088–5095.

Cope L D, Lumbley S, Latimer J L, Klesney-Tait J, Stevens M K, Johnson L S, Purven M, Munson R S Jr, Lagergard T, Radolf J D, and Hansen E J (1997). A diffusible cytotoxin of *Haemophilus ducreyi*. *Proc. Natl. Acad. Sci. USA*, **94**, 4056–4061.

Cortes-Bratti X, Frisan T, and Thelestam M (2001a). The cytolethal distending toxins induce DNA damage and cell cycle arrest. *Toxicon*, **39**, 1729–1736.

Cortes-Bratti X, Karlsson C, Lagergard T, Thelestam M, and Frisan T (2001b). The *Haemophilus ducreyi* cytolethal distending toxin induces cell cycle arrest and apoptosis via the DNA damage checkpoint pathways. *J. Biol. Chem.*, **276**, 5296–5302.

Cossart P, Boquet P, Normark S, and Rappuoli R (1996). Cellular microbiology emerging. *Science*, **271**, 315–316.

De Rycke J, Mazars P, Nougayrede J P, Tasca C, Boury M, Herault F, Valette A, and Oswald E (1996). Mitotic block and delayed lethality in HeLa epithelial cells exposed to *Escherichia coli* BM2–1 producing cytotoxic necrotizing factor type 1. *Infect. Immun.*, **64**, 1694–1705.

De Rycke J and Oswald E (2001). Cytolethal distending toxin (CDT): A bacterial weapon to control host cell proliferation? *FEMS Microbiol. Lett.*, **203**, 141–148.

De Rycke J, Sert V, Comayras C, and Tasca C (2000). Sequence of lethal events in HeLa cells exposed to the G2 blocking cytolethal distending toxin. *Eur. J. Cell Biol.*, **79**, 192–201.

Demuth D R, Savary R, Golub E, and Shenker B J (1996). Identification and analysis of fipA, a *Fusobacterium nucleatum* immunosuppressive factor gene. *Infect. Immun.*, **64**, 1335–1341.

Deng K, Latimer J L, Lewis D A, and Hansen E J (2001). Investigation of the interaction among the components of the cytolethal distending toxin of *Haemophilus ducreyi*. *Biochem. Biophys. Res. Commun.*, **285**, 609–615.

Denko N, Langland R, Barton M, and Lieberman M A (1997). Uncoupling of S-phase and mitosis by recombinant cytotoxic necrotizing factor 2 (CNF2). *Exp. Cell Res.*, **234**, 132–138.

Elwell C, Chao K, Patel K, and Dreyfus L (2001). *Escherichia coli* CdtB mediates cytolethal distending toxin cell cycle arrest. *Infect. Immun.*, **69**, 3418–3422.

Elwell C A and Dreyfus L A (2000). DNase I homologous residues in CdtB are critical for cytolethal distending toxin-mediated cell cycle arrest. *Mol. Microbiol.*, **37**, 952–963.

Escalas N, Davezac N, De Rycke J, Baldin V, Mazars R, and Ducommun B (2000). Study of the cytolethal distending toxin-induced cell cycle arrest in HeLa cells: Involvement of the CDC25 phosphatase. *Exp. Cell Res.*, **257**, 206–212.

Frisan T, Cortes-Bratti X, Chaves-Olarte E, Stenerlow B, and Thelestam M (2003). The *Haemophilus ducreyi* cytolethal distending toxin induces DNA double-strand breaks and promotes ATM-dependent activation of RhoA. *Cell Microbiol.*, **5**, 695–707.

Frisk A, Lebens M, Johansson C, Ahmed H, Svensson L, Ahlman K, and Lagergard T (2001). The role of different protein components from the *Haemophilus ducreyi* cytolethal distending toxin in the generation of cell toxicity. *Microb. Pathogenesis.*, **30**, 313–324.

Gachet Y, Tournier S, Millar J B, and Hyams J S (2001). A MAP kinase-dependent actin checkpoint ensures proper spindle orientation in fission yeast. *Nature*, **412**, 352–355.

Gelfanova V, Hansen E J, and Spinola S M (1999). Cytolethal distending toxin of *Haemophilus ducreyi* induces apoptotic death of Jurkat T cells. *Infect. Immun.*, **67**, 6394–6402.

Gutkind J S (1998). The pathways connecting G protein-coupled receptors to the nucleus through divergent mitogen-activated protein kinase cascades. *J. Biol. Chem.*, **273**, 1839–1842.

Harrison J C, Bardes E S, Ohya Y, and Lew D J (2001). A role for the Pkc1p/Mpk1p kinase cascade in the morphogenesis checkpoint. *Nat. Cell Biol.*, **3**, 417–420.

Hartwell L H (1978). Cell division from a genetic perspective. *J. Cell Biol.*, **77**, 627–637.

Hassane D C, Lee R B, Mendenhall M D, and Pickett C L (2001). Cytolethal distending toxin demonstrates genotoxic activity in a yeast model. *Infect. Immun.*, **69**, 5752–5759.

Henderson B, Wilson M, and Hyams J (1998). Cellular microbiology: Cycling into the millennium. *Trends Cell Biol.*, **8**, 384–387.

Horiguchi Y (2001). *Escherichia coli* cytotoxic necrotizing factors and *Bordetella* dermonecrotic toxin: The dermonecrosis-inducing toxins activating Rho small GTPases. *Toxicon*, **39**, 1619–1627.

Hudson K M, Denko N C, Schwab E, Oswald E, Weiss A, and Lieberman M A (1996). Megakaryocytic cell line-specific hyperploidy by cytotoxic necrotizing factor bacterial toxins. *Blood*, **88**, 3465–3473.

Johnson W M and Lior H (1988). A new heat-labile cytolethal distending toxin (CLDT) produced by *Campylobacter* spp. *Microb. Pathogenesis.*, **4**, 115–126.

Kaszkin M, Richards J, and Kinzel V (1996). Phosphatidic acid mobilized by phospholipase D is involved in the phorbol 12-myristate 13-acetate-induced G2 delay of A431 cells. *Biochem. J.*, **314**, 129–138.

King R, Gough J, Ronald A, Nasio J, Ndinya-Achola J O, Plummer F, and Wilkins J A (1996). An immunohistochemical analysis of naturally occurring chancroid. *J. Infect. Dis.*, **174**, 427–430.

Kinzel V, Kaszkin M, Blume A, and Richards J (1990). Epidermal growth factor inhibits transiently the progression from G2- phase to mitosis: A receptor-mediated phenomenon in various cells. *Cancer Res.*, **50**, 7932–7936.

Klapproth J M, Scaletsky I C, McNamara B P, Lai L C, Malstrom C, James S P, and Donnenberg M S (2000). A large toxin from pathogenic *Escherichia coli* strains that inhibits lymphocyte activation. *Infect. Immun.*, **68**, 2148–2155.

Knipp U, Birkholz S, Kaup W, and Opferkuch W (1996). Partial characterization of a cell proliferation-inhibiting protein produced by *Helicobacter pylori*. *Infect. Immun.*, **64**, 3491–3496.

Lara-Tejero M and Galan J E (2000). A bacterial toxin that controls cell cycle progression as a deoxyribonuclease I-like protein. *Science*, **290**, 354–357.

Lara-Tejero M and Galan J E (2001). CdtA, CdtB, and CdtC form a tripartite complex that is required for cytolethal distending toxin activity. *Infect. Immun.*, **69**, 4358–4365.

Lara-Tejero M and Galan J E (2002). Cytolethal distending toxin: Limited damage as a strategy to modulate cellular functions. *Trends Microbiol.*, **10**, 147–152.

Lewis D A, Stevens M K, Latimer J L, Ward C K, Deng K, Blick R, Lumbley S R, Ison C A, and Hansen E J (2001). Characterization of *Haemophilus ducreyi* cdtA, cdtB, and cdtC mutants in in vitro and in vivo systems. *Infect. Immun.*, **69**, 5626–5634.

Li L, Sharipo A, Chaves-Olarte E, Masucci M G, Levitsky V, Thelestam M, and Frisan T (2002). The *Haemophilus ducreyi* cytolethal distending toxin activates sensors of DNA damage and repair complexes in proliferating and non-proliferating cells. *Cell Microbiol.*, **4**, 87–99.

Mailand N, Falck J, Lukas C, Syljuasen R G, Welcker M, Bartek J, and Lukas J (2000). Rapid destruction of human Cdc25A in response to DNA damage. *Science*, **288**, 1425–1429.

Mao X and DiRienzo J M (2002). Functional studies of the recombinant subunits of a cytolethal distending holotoxin. *Cell. Microbiol.*, **4**, 245–255.

Marches O, Ledger T N, Boury M, Ohara M, Tu X, Goffaux F, Mainil J, Rosenshine I, Sugai M, De Rycke J, and Oswald E (2003). Enteropathogenic and enterohaemorrhagic *Escherichia coli* deliver a novel effector called Cif, which blocks cell cycle G2/M transition. *Mol. Microbiol.*, **50**, 1553–1567.

Matsui K, Nagano K, Arai T, Hirono I, and Aoki T (1998). DNA sequencing of the gene encoding *Salmonella typhimurium*-derived T-cell inhibitor (STI) and characterization of the gene product, cloned STI. *FEMS Immunol. Med. Mic.*, **22**, 341–349.

Morgan D O (1999). Regulation of the APC and the exit from mitosis. *Nat. Cell Biol.*, **1**, 47–53.

Nilsson I and Hoffmann I (2000). Cell cycle regulation by the Cdc25 phosphatase family. *Prog. Cell Cycle Res.*, **4**, 107–114.

Nougayrede J P, Marches O, Boury M, Mainil J, Charlier G, Pohl P, De Rycke J, Milon A, and Oswald E (1999). The long-term cytoskeletal rearrangement induced by rabbit enteropathogenic *Escherichia coli* is Esp dependent but intimin independent. *Mol. Microbiol.*, **31**, 19–30.

Nurse P (1990). Universal control mechanism regulating onset of M-phase. *Nature*, **344**, 503–508.

O'Connell M J, Walworth N C, and Carr A M (2000). The G2-phase DNA-damage checkpoint. *Trends Cell Biol.*, **10**, 296–303.

Okuda J, Fukumoto M, Takeda Y, and Nishibuchi M (1997). Examination of diarrheagenicity of cytolethal distending toxin: Suckling mouse response to the products of the cdtABC genes of *Shigella dysenteriae*. *Infect. Immun.*, **65**, 428–433.

Op De Beeck A and Caillet-Fauquet P (1997). Viruses and the cell cycle. *Prog. Cell Cycle Res.*, **3**, 1–19.

Peres S Y, Marches O, Daigle F, Nougayrede J P, Herault F, Tasca C, De Rycke J, and Oswald E (1997). A new cytolethal distending toxin (CDT) from *Escherichia coli* producing CNF2 blocks HeLa cell division in G2/M phase. *Mol. Microbiol.*, **24**, 1095–1107.

Pickett C L, Cottle D L, Pesci E C, and Bikah G (1994). Cloning, sequencing, and expression of the *Escherichia coli* cytolethal distending toxin genes. *Infect. Immun.*, **62**, 1046–1051.

Pines J (1995). Cyclins and cyclin-dependent kinases: A biochemical view. *Biochem. J.*, **308**, 697–711.

Pitari G M, Di Guglielmo M D, Park J, Schulz S, and Waldman S A (2001). Guanylyl cyclase C agonists regulate progression through the cell cycle of human colon carcinoma cells. *Proc. Natl. Acad. Sci. USA*, **98**, 7846–7851.

Prepens U, Barth H, Wilting J, and Aktories K (1998). Influence of *Clostridium botulinum* C2 toxin on Fc epsilonRI-mediated secretion and tyrosine phosphorylation in RBL cells. *N-Sc Arch. Pharmacol.*, **357**, 323–330.

Purdy D, Buswell C M, Hodgson A E, McAlpine K, Henderson I, and Leach S A (2000). Characterisation of cytolethal distending toxin (CDT) mutants of *Campylobacter jejuni*. *J. Med. Microbiol.*, **49**, 473–479.

Rhind N and Russell P (2001). Roles of the mitotic inhibitors Wee1 and Mik1 in the G(2) DNA damage and replication checkpoints. *Mol. Cell Biol.*, **21**, 1499–1508.

Ridley A J (1995). Rho-related proteins: Actin cytoskeleton and cell cycle. *Curr. Opin. Genet. Dev.*, **5**, 24–30.

Robinson D N and Spudich J A (2000). Towards a molecular understanding of cytokinesis. *Trends Cell Biol.*, **10**, 228–237.

Saiki K, Konishi K, Gomi T, Nishihara T, and Yoshikawa M (2001). Reconstitution and purification of cytolethal distending toxin of *Actinobacillus actinomycetemcomitans*. *Microbiol. Immunol.*, **45**, 497–506.

Scott D A and Kaper J B (1994). Cloning and sequencing of the genes encoding *Escherichia coli* cytolethal distending toxin. *Infect. Immun.*, **62**, 244–251.

Sert V, Cans C, Tasca C, Bret-Bennis L, Oswald E, Ducommun B, and De Rycke J (1999). The bacterial cytolethal distending toxin (CDT) triggers G2 cell cycle checkpoint in mammalian cells without preliminary induction of DNA strand breaks. *Oncogene*, **18**, 6296–6304.

Shenker B J and Datar S (1995). *Fusobacterium nucleatum* inhibits human T-cell activation by arresting cells in the mid-G1 phase of the cell cycle. *Infect. Immun.*, **63**, 4830–4836.

Shenker B J, Hoffmaster R H, McKay T L, and Demuth D R (2000). Expression of the cytolethal distending toxin (Cdt) operon in *Actinobacillus actinomycetemcomitans*: Evidence that the CdtB protein is responsible for G2 arrest of the cell cycle in human T cells. *J. Immunol.*, **165**, 2612–2618.

Shenker B J, Hoffmaster R H, Zekavat A, Yamaguchi N, Lally E T, and Demuth D R (2001). Induction of apoptosis in human T cells by *Actinobacillus actinomycetemcomitans* cytolethal distending toxin is a consequence of G2 arrest of the cell cycle. *J. Immunol.*, **167**, 435–441.

Shenker B J, McKay T, Datar S, Miller M, Chowhan R, and Demuth D (1999). *Actinobacillus actinomycetemcomitans* immunosuppressive protein is a member of the family of cytolethal distending toxins capable of causing a G2 arrest in human T cells. *J. Immunol.*, **162**, 4773–4780.

Shenker B J, Vitale L, and Slots J (1991). Immunosuppressive effects of *Prevotella intermedia* on in vitro human lymphocyte activation. *Infect. Immun.*, **59**, 4583–4589.

Sherr C J (1994). The ins and outs of RB: Coupling gene expression to the cell cycle clock. *Trends Cell Biol.*, **4**, 15–19.

Sherr C J and Roberts J M (1999). CDK inhibitors: Positive and negative regulators of G1-phase progression. *Gene Dev.*, **13**, 1501–1512.

Stevens M K, Latimer J L, Lumbley S R, Ward C K, Cope L D, Lagergard T, and Hansen E J (1999). Characterization of a *Haemophilus ducreyi* mutant deficient in expression of cytolethal distending toxin. *Infect. Immun.*, **67**, 3900–3908.

Sugai M, Kawamoto T, Peres S Y, Ueno Y, Komatsuzawa H, Fujiwara T, Kurihara H, Suginaka H, and Oswald E (1998). The cell cycle-specific growth-inhibitory factor produced by *Actinobacillus actinomycetemcomitans* is a cytolethal distending toxin. *Infect. Immun.*, **66**, 5008–5019.

Svensson L A, Tarkowski A, Thelestam M, and Lagergard T (2001). The impact of *Haemophilus ducreyi* cytolethal distending toxin on cells involved in immune response. *Microb. Pathogenesis.*, **30**, 157–166.

Tan M, Jing T, Lan K H, Neal C L, Li P, Lee S, Fang D, Nagata Y, Liu J, Arlinghaus R, Hung M C, and Yu D H (2002). Phosphorylation on tyrosine-15 of p34(Cdc2) by ErbB2 inhibits p34(Cdc2) activation and is involved in resistance to taxol-induced apoptosis. *Mol. Cell.*, **9**, 993–1004.

van Setten P A, van Hinsbergh V W, van den Heuvel L P, van der Velden T J, van de Kar N C, Krebbers R J, Karmali M A, and Monnens L A (1997). Verocytotoxin inhibits mitogenesis and protein synthesis in purified human glomerular mesangial cells without affecting cell viability: Evidence for two distinct mechanisms. *J. Am. Soc. Nephrol.*, **8**, 1877–1888.

Walworth N C (2001). DNA damage: Chk1 and Cdc25, more than meets the eye. *Curr. Opin. Genet. Dev.*, **11**, 78–82.

Wilson M and Henderson B (1995). Virulence factors of *Actinobacillus actinomycetemcomitans* relevant to the pathogenesis of inflammatory periodontal diseases. *FEMS Microbiol. Rev.*, **17**, 365–379.

Yoshida J, Ishibashi T, and Nishio M (2001). Growth-inhibitory effect of a streptococcal antitumor glycoprotein on human epidermoid carcinoma A431 cells: Involvement of dephosphorylation of epidermal growth factor receptor. *Cancer Res.*, **61**, 6151–6157.

Yoshida J, Takamura S, and Nishio M (1998). Characterization of a streptococcal antitumor glycoprotein (SAGP). *Life Sci.*, **62**, 1043–1053.

Young R S, Fortney K R, Gelfanova V, Phillips C L, Katz B P, Hood A F, Latimer J L, Munson R S Jr, Hansen E J, and Spinola S M (2001). Expression of cytolethal distending toxin and hemolysin is not required for pustule formation by *Haemophilus ducreyi* in human volunteers. *Infect. Immun.*, **69**, 1938–1942.

Zhou B B and Elledge S J (2000). The DNA damage response: Putting checkpoints in perspective. *Nature*, **408**, 433–439.

CHAPTER 5

Bartonella signaling and endothelial cell proliferation

Garret Ihler, Anita Verma, and Javier Arevalo

Bartonella bacilliformis is the causative agent of Carrion's disease (for a recent medical review, see Maguina and Gotuzzo, 2000). *B. bacilliformis* was in a genus by itself until recent taxonomic revisions associated it with other bacteria. Now there are sixteen known species of *Bartonella*, of which seven are associated with human disease. *B. bacilliformis* enters the bloodstream of humans through the bite of an insect vector and targets primarily erythrocytes and endothelial cells.

For many years, it seemed as though important and interesting conclusions about the mechanism of angiogenesis might result from the study of *B. bacilliformis*, which can initiate and perpetuate lesions formed of proliferating endothelial cells and newly formed capillaries. Since then, the Science juggernaut has moved on, more rapidly for angiogenesis than for *Bartonella*, and so it now appears that we can best understand *B. bacilliformis* (and the closely related *Bartonella henselae*) by reference to the known biology of angiogenesis. However, study of *B. bacilliformis* has made at least one useful contribution to our knowledge of angiogenesis. Entry of *B. bacilliformis* by vacuole formation and macropinocytosis was shown to be accomplished by activation of Rho, Rac, and Cdc42 GTPases. This process appeared to be analogous to the angiogenic process by which endothelial cells take in large volumes of extracellular fluids to form large central vacuoles, which perhaps fuse between cells to form the nascent capillary. Following the studies with *B. bacilliformis*, it was shown that this process of endothelial cell fluid intake is accomplished by activation of Rac and Cdc42 (Bayless and Davis, 2002).

Bartonella spp. have progressed in recent years from being totally obscure pathogens to ones with relatively few infected patients and relatively many review articles, some of which are listed here (Ihler, 1996; Dehio, 1999;

Breitschwerdt and Kordick, 2000; Karem, 2000; Karem et al., 2000; Minnick and Anderson, 2000; Dehio, 2001; Houpikian and Raoult, 2001).

UNDERSTANDING DISEASE CAUSED BY *BARTONELLA*

We believe the most important guiding principle at present for understanding *Bartonella* is that the events after infection of endothelial cells by the bacteria are mediated by manipulating host cell programs. The consequences of infection therefore should be reminiscent of what is known to happen in ordinary angiogenesis or tumor-associated angiogenesis, although the results of infection could differ in degree because multiple programs could be activated to greater or lesser extent than observed in normal angiogenesis. A corollary is that it will not be possible to understand all the details of *B. bacilliformis*–promoted angiogenesis from simply studying bacterial infection of endothelial cells *in vitro* because angiogenesis involves cell signaling and the interactions of several cell types, and is dependent on the interstitial matrix.

An additional guiding principle is attention to the life cycle of *B. bacilliformis* and *B. henselae*. The bacteria are transmitted from a blood-sucking insect vector into the blood of a vertebrate host and establish protected sites in the host, with access to the blood, where the bacteria may persist long enough to allow re-infection of other blood-sucking insects. In humans, *B. bacilliformis*–invaded erythrocytes provide a protected site, with a potential life span of 120 days. *B. henselae* is endemic in cats, and has been reported both to be present (Kordick and Breitschwerdt, 1995; Rolain et al., 2001) and absent from cat erythrocytes (Guptill et al., 2000), but in any case does not extensively invade human erythrocytes (Iwaki-Egawa and Ihler, 1997). Life cycle considerations suggest that the relative lack of ability to infect human erythrocytes may explain why humans are only incidental hosts for *B. henselae*, since *B. henselae* usually will not be retransmitted to the vector from humans. However, it should be noted that about 20% of patients with cat scratch disease who were serologically positive for *Bartonella* had bacteremia as shown by PCR. The PCR-amplified sequence was identical to that of *B. henselae* (Tsukahara et al., 2001). Additional protected sites are tumor-like lesions in skin caused to form by the bacteria, called in humans verruga peruana (*B. bacilliformis*) and bacillary angiomatosis (*B. henselae*). These lesions contain bacteria and are composed of proliferating endothelial cells and capillaries. From these sites, the bacteria have direct access to the blood. The long life span of the verruga (more than a year) suggests that *B. bacilliformis* can manipulate the immune response. Infection by *B. henselae* in an immunocompetent host is eventually controlled ("cat-scratch fever") and in an immunocompromised host, usually an AIDS patient, results in a lesion very similar to that of *B. bacilliformis*. In humans, *B. bacilliformis* can render ineffective some aspect of the human immune response which *B. henselae* cannot. An interesting case history of a patient with rheumatoid arthritis treated with corticosteroids and methotrexate who became infected with *B. henselae* suggests caution in cat ownership for the immunosuppressed (Harsch et al., 2002).

B. BACILLIFORMIS INFECTION INDUCES ABERRANT ANGIOGENESIS

There are several sizes and forms of verrugal lesions. The reasons for the variations are not clear, but may relate to where the lesion is located and whether it is in the papillary dermis (miliary form), the subcutaneous tissue, or involving all layers of dermis. Larger lesions are located on the arms and legs (nodular) or appear near joints (mular or subdermic nodules). Deep-seated nodules tend to be compact (Arias-Stella et al., 1986).

Microscopically, lesions consist of a network of capillaries in association with endothelial hyperplasia and dividing endothelial cells. In some cases sheets of cells in a pseudoepitheloid pattern are seen, in which the capillaries are collapsed; in other cases well-established capillaries with open lumens are observed. Predominantly spindle-shaped cells were seen (Shen et al., 1998). Dense infiltrations of lymphocytes, monocytes, macrophages, and dendocytic cells are found, indicating active inflammation. Bacteria are seen in extracellular spaces near blood vessels and in intracellular locations within endothelial cells. Despite the evidence of active inflammation, the verrugal lesions can persist as long as 12–18 months. Bacteremia may continue, sometimes intermittently, for about the same period of time, about 15 months. Presumably, as long as the verrugas persist, bacteremia and transfer to the insect vector is possible, since the verrugas serve as the reservoir.

Likewise, the cells of bacillary angiomatosis (BA) are activated endothelial cells, with numerous Weibel-Palade bodies, and express a variety of endothelial markers. Numerous cells expressing macrophage/monocyte markers are present in BA surrounding and within the lesions. In many respects, the lesions of both *B. bacilliformis* and *B. henselae* closely resemble those of Kaposi's sarcoma, and numerous analogies can be profitably drawn, despite the fact that the etiological agent in Kaposi's sarcoma is a virus – Kaposi's sarcoma-associated herpes virus. The bacterial lesions can be distinguished from those in Kaposi's sarcoma by surface immunohistochemistry of the endothelial cells, by the presence of neutrophils, and of course by the absence of bacteria in Kaposi's sarcoma (LeBoit et al., 1989; Kostianovsky et al., 1992). These angiogenic lesions also have similarities with hemangiomas or pyogenic granulomas. A possible association between *Bartonella* and pyogenic granulomas – benign vascular lesions – has been discussed (Lee and Lynde, 2001). Verrugas are primarily in the skin, whereas BA can occur in various organs and can have unexpected presentations, including the eye (neuroretinitis with macular edema and exudates (Earhart and Power, 2000)). Peliosis can be caused by *B. henselae*, and bacilli are found in the hepatic sinusoidal endothelial cells, both intracellularly and extracellularly, and in associated abdominal lymph nodes.

RECOGNITION BETWEEN *B. BACILLIFORMIS* AND ERYTHROCYTES OR ENDOTHELIAL CELLS

Binding of bacteria both to erythrocytes and endothelial cells relies on the bacterial outer membrane. Omp43, an outer membrane protein (related to Omp2b of *Brucella*) has been shown to bind to endothelial cells (Burgess et al., 2000). *B. bacilliformis* has been shown to bind several biotinylated erythrocyte surface proteins (Buckles and McGinnis Hill, 2000). Between 1991 and 1993, Javier Arevalo investigated the interactions between outer membrane proteins from *B. bacilliformis* and erythrocytes, BL3 cells and HUVECs. Because these experiments have not been published, we here summarize the results, which are available with the data on the Department of Medical Biochemistry and Genetics web site at Texas A&M College of Medicine (http://128.194.251.107/Ihler_home_page/).

1. Complete antibodies generated against the sarkosyl-insoluble membrane proteins of *B. bacilliformis* reduced binding of the bacteria to erythrocytes up to 90%. Fab fragments, which could be used without aggregating the bacteria, completely prevented binding.
2. The outer membrane proteins of *B. bacilliformis* were characterized. Sarkosyl-insoluble proteins from the membrane fraction of French pressed cells or membrane proteins prepared by the lithium acetate procedure were designated as outer membrane proteins after comparison with outer membrane proteins separated on step sucrose gradients. Detergent-solubilized proteins could not be used for binding studies with erythrocytes because the detergent lysed the erythrocytes, even when using the lowest possible concentrations of several nonionic detergents. It was found the supernatant from cultures contained most

of the proteins present in membrane fractions, including all the sarkosyl-insoluble proteins, in a soluble or at least a dispersed state, and these could be radioactively labeled with ^{35}S-methionine by growing the bacteria in Delbecco's modified Earle's medium deficient in methionine. Species of protein that bound to erythrocytes were 95-, 80-, and 52-kDa with lesser amounts of 65- and 45-kDa species. The 95-kDa protein was soluble in 1% sarkosyl. All these could be released by increasing the salt concentration, suggesting they interact by ionic forces. *Escherichia coli* proteins did not bind.

3. Biotinylated erythrocyte membrane proteins were incubated with sarkosyl-insoluble proteins or lithium acetate–prepared proteins, which had been separated by SDS-PAGE and transferred to nitrocellulose. The erythrocyte membrane proteins bound to 80-, 45-, and 38-kDa species of *B. bacilliformis* sarkosyl-insoluble proteins, but did not recognize *E. coli* proteins. Erythrocyte proteins recognized the 95-, 80-, 45-, and 38-kDa species of lithium acetate-prepared proteins, and in addition proteins at 55-, 49-, and 24-kDa. The reverse experiment, in which erythrocyte membrane proteins were separated by SDS-PAGE, was not successful, probably because the ability to bind was destroyed by denaturation. That the *B. bacilliformis* proteins could be electrophoresed in SDS and retained activity suggests that they were not irreversibly affected, perhaps because of their hydrophobic properties. When the total protein complement of *B. bacilliformis* was separated by SDS-PAGE, only outer membrane proteins were bound by biotinylated erythrocyte proteins.

4. Affinity column chromatography using Affigel-immobilized sarkosyl-insoluble proteins identified an 18.5-kDa erythrocyte membrane protein that could be released from the column with 1.5 M NaCl, and an additional 4 erythrocyte proteins between 50–65-kDa that could be released only by boiling in sample buffer.

5. Membrane proteins from BL-3 and HUVEC cells, but not erythrocytes, separated by SDS-PAGE could be recognized in Western blots by radioactive intact *B. bacilliformis*.

6. The flagellar gene product, *fla*A, which was cloned, sequenced, and expressed, does not bind to erythrocytes.

Porins and Blebbing

Not much is known about the role of *Bartonella* porins or outer membrane proteins, specifically for pathogenesis. Porins are likely to be of great importance. For example, *Neisseria*, which has similarities to *Bartonella*, can utilize PorB to modulate apoptosis, because PorB is efficiently targeted to mitochondria (Muller et al., 2002), where it can depolarize the mitochondrion. Like *Neisseria*, *Bartonella* produces membranous blebs containing some but not all of the outer membrane proteins (J. Arevalo and G. Ihler, unpublished). These blebs have considerable pathogenic potential, since they can distribute the outer membrane proteins widely and independently of viable bacteria. Not only could these serve as decoys or as sinks for host antibacterial proteins, but also if the proteins themselves have toxic properties, the blebs could deliver them to host cells as efficiently as the live bacteria. Neisserial porins can stimulate B cells and upregulate surface expression of host proteins via an NF-κB–dependent mechanism (Massari et al., 2002).

Why Are Erythrocytes and Endothelial Cells the Two Primary Targets for *B. bacilliformis*?

Bartonella are the only bacteria known to invade human erythrocytes. Infection by *B. bacilliformis* first involves colonization of the erythrocytes (Figure 5.1), massive destruction of erythrocytes by spleen and liver, and a resulting severe anemia, which is often fatal in the absence of antibiotic treatment (Oroya fever). Invasion of endothelial cells and subsequent angiogenesis results in the formation of verrugas. Verruga peruana has low or zero mortality but involves formation of the numerous disfiguring skin lesions. The net result is a two-phase disease (Carrion's disease), although in some cases the erythrocytic phase is inapparent.

It seems likely that there must be some underlying connection between these two phases of the disease and the two cell types. But it does not seem likely that *B. bacilliformis* uses the same mechanism of invasion for erythrocytes and nucleated cells because mature erythrocytes do not engage in endocytosis or vesicle transport. Both infection and retransmission must

Figure 5.1 (Bacterial morphology and interaction with erythrocytes A. Colony morphology at 14 days. Left: 440× Right: 4400×. B. Colony face split open. 9600× (A,B: G. McLaughlin and G. Ihler, unpublished). C. Bacterial clump adherent to a red cell. 10,100×. D. Adherent bacteria with small indentation. Individual bacteria are often seen in deep invaginations of the erythrocyte membrane (Benson et al., 1986). E–F. Bacteria associated with trenches and pits. The lips of the trench sometimes seem to fold over, as if engulfing the bacteria. Possibly this could occur if the clump binds to two or more sites on the erythrocyte surface as in Figure 1C and then contracts to pull the erythrocyte membrane over the clump.

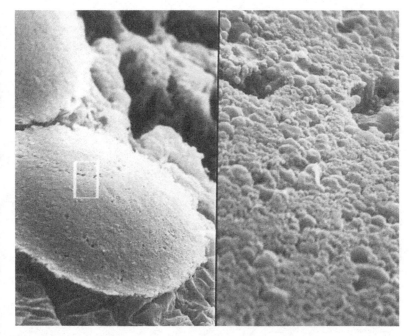

Figure 5.1A

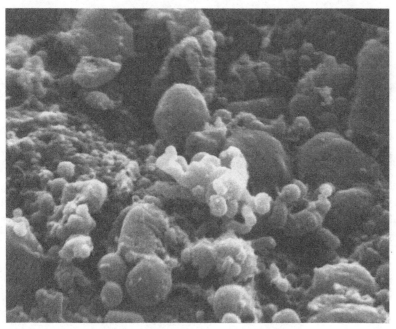

Figure 5.1B

(Continued)

Figure 5.1C

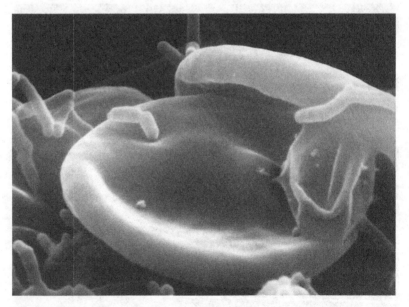

Figure 5.1D

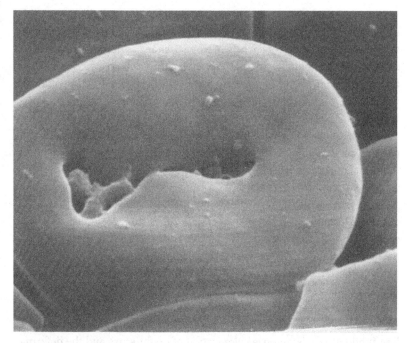

Figure 5.1E

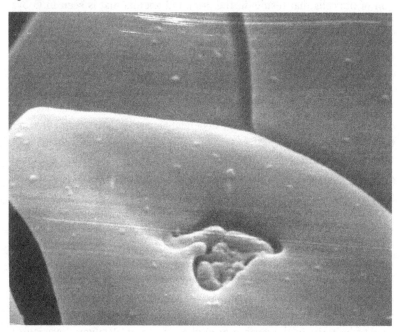

Figure 5.1F

involve the blood, so it is easy to see why attachment to erythrocytes might be advantageous. The intense monitoring of erythrocytes in humans by the spleen might be part of the reason why a tissue sanctuary is a feature of the disease, and the long life span of the verruga (more than a year) relative to the maximum life span of an erythrocyte (120 days) could be important as well. *B. henselae* manages to cause formation in humans of lesions similar to verrugas without a prominent erythrocytic phase, but there is no evidence that *B. henselae* can be re-transmitted to an insect vector from humans (although patients are certainly bacteremic). *Bartonella tribocorum*, a rat pathogen, invades and replicates in erythrocytes, but the erythrocytes are not markedly cleared (Schulein et al., 2001).

Although *B. bacilliformis* can be internalized into erythrocytes and circulate protected within the erythrocytes, it does not accomplish this in a stealthy manner. Instead, clumps of bacteria adhere to the external surface of the erythrocytes, where they are exposed to antibodies and the immune system. The bacteria induce major deformations in the erythrocyte membrane, including pits and trenches (Benson et al., 1986; Derrick and Ihler, 2001). These changes cause erythrocytes to be recognized as defective, especially in the splenic circulation, but also by liver macrophages if the damage is less subtle. It is this massive colonization of erythrocytes and the deformation of the cells that results in the profound anemia that is seen in Oroya fever because erythrocytes deformed by *B. bacilliformis* infection are readily phagocytosed.

Phagocytosis of erythrocytes by macrophages might suggest that phosphatidylserine – a signal for phagocytosis not only in erythrocytes, but in most cells – is exposed on the outer face of the erythrocyte membrane. In normal erythrocytes, phosphatidylserine is absent from the external face because it is transported by an energy-dependent enzyme system from the outer to the inner leaflet of the membrane bilayer. Exposure of phosphatidylserine endows erythrocytes with the propensity of adhering to HUVECs (Closse et al., 1999) as well as to phagocytic cells, although normal erythrocytes are generally non-adhesive to endothelial surfaces.

Endothelial cells can be readily infected *in vitro* by free bacteria, and presumably also within the verruga. But the initial colonization at the site of a future verruga might be facilitated if the bacteria are bound to and circulating with erythrocytes, either because the erythrocyte is phagocytosed or because the bacteria are brought into close proximity to the endothelial cell. An erythrocytic route is suggested by the fact that erythrocyte colonization constitutes the first phase of the disease. One might speculate that the presence of bacteria on the surface of erythrocytes could facilitate invasion of endothelial cells as a second protected site because erythrocytes fill small capillaries and are deformed by their passage. Although endothelial cells are sometimes described as wrapping around or lining capillaries, in fact the capillary has no physical existence except as a hole through endothelial cells. Thus the bacteria would "scrape" along the endothelial surface and could perhaps attach more easily to the endothelial cells after being delivered by carrier erythrocytes.

Transfer of bacteria to endothelial cells could be additionally facilitated if the erythrocytes or the bacteria that were bound to the erythrocytes adhered to the capillary endothelium; reverse transfer of bacteria from the verrugas to erythrocytes might also be possible if erythrocytes became arrested in capillaries or post-capillary venules. In sickle-cell anemia, adhesion pathways have been identified as contributing to the sickle-cell crisis, in addition to, or as a consequence of, erythrocyte deformation caused by HbS polymerization. It is noteworthy that the sickle-cell crisis is very often associated with infection. One possibility might be that infection systemically upregulates adhesive proteins on endothelial cells, which become more effective in arresting abnormal sickle-cell erythrocytes.

The importance of cell adhesion molecules in leukocyte trafficking and in mediating immune and inflammatory responses ensures that cell–cell recognition proteins would certainly be upregulated in verrugas, as in other infections, but it is not known whether these adhesive proteins would help to bind erythrocytes. *B. henselae* outer membrane proteins (and perhaps LPS) are known to upregulate E-selectin and ICAM-1 on HUVECs (Fuhrmann et al., 2001; Maeno et al., 2002). For ICAM-1, infection is not required, since the nonpiliated strain, unable to invade endothelial cells, induced ICAM-1 as well as the piliated strain, and since the bacteria could be inactivated in various ways (including being sonicated) without loss of activity. The unidentified factor was stable to boiling and unaffected by the addition of polymyxin B, suggesting that it was not LPS.

B. bacilliformis and *B. henselae* Infect Endothelial Cells

Invasion of endothelial cells by *B. bacilliformis* is considered to be a crucial and important step in establishing the proliferative lesions of the verruga. Invasion of endothelial cells by different species of *Bartonella* has been studied primarily *in vitro* using human umbilical vein endothelial cells (HUVECs).

B. bacilliformis infections *in vivo* are characterized by bacteria within and outside endothelial cells. *Bartonella* is tropic for endothelial cells, and has special growth effects on them. The significance of this tropism is somewhat confused by the fact that *Bartonella* can infect many cell types *in vitro*, in addition to endothelial cells, although growth of these cells is not stimulated.

Thus, if a specialized mechanism of *Bartonella* entry into endothelial cells exists, *Bartonella* must possess some additional, more general mechanism for gaining access to different types of cells. Despite this caveat, it nevertheless does appear that *Bartonella* has a special tropism for endothelial cells, and a special mechanism of entry.

In vitro, intracellular *B. bacilliformis* were observed electron microscopically in small membrane-bound inclusions within endothelial cells after 1 h of infection and by 12 h a large membrane-bound inclusion containing numerous bacteria was present. This was described as being similar to the Rocha-Lima inclusion, which is a prominent feature of *in vivo* infections (Garcia et al., 1992). Cytochalasin-D reduced the invasiveness of *B. bacilliformis* into endothelial cells *in vitro* (Hill et al., 1992), indicating host participation.

Activation of Rho-family Small GTPases

Verma et al. (2000, 2001) showed that a significant entry of *B. bacilliformis* occurs within 2 hrs, and also that there is also a very large increase in entry between 16 and 24 h. A massive reorganization of the actin network leading to the formation of thick actin bundles, known as stress fibers, was visible in the infected endothelial cells by 12–16 h. It has been shown that a member of the small GTPase family of proteins, Rho, is a key signaling molecule for mediating the formation of stress fibers. Members of the Rho family of small GTP-binding proteins (Rho, Rac, and Cdc42) function as molecular switches by cycling between an inactive state with bound GDP and an active state with bound GTP (see Chapter 3 for more details). The active form of Rho interacts with downstream effector proteins to produce biological responses, which include actin reorganization. Rho is also necessary for internalization of *B. bacilliformis*. Entry at both early and late times can be prevented by pretreatment of the endothelial cells with C3 exoenzyme, a protein toxin from *Clostridium botulinum* which inactivates intracellular Rho by ADP ribosylation. Moreover, Rho was shown to be directly activated in the infected HUVECs, with a similar time course to that observed for bacterial invasion. In a later study, Verma and Ihler (2002) also showed that pretreatment with the *Clostridial* toxin TcsL-1522 – which specifically inactivates another member of the small GTP-binding protein family, Rac and, to a lesser extent, Cdc42, but not Rho – also inhibits entry. All three Rho family proteins, Rac, Cdc42, and Rho, are activated after incubation of endothelial cells with *B. bacilliformis*. Within 30 min, levels of activated Rac and activated Cdc42 were increased.

Relocalization of other Rho family proteins also occurs. Within 30 min, filopodia (filamentous actin extensions) formed; within 1 h lamellipodia (membrane rufflings) formed. Clumps of bacteria could be seen adhering to or in the close vicinity of both the filopodia and lamellipodia. F-actin was associated with both filopodia and lamellipodia, and Rac was shown to be associated with the lamellipodia (Verma and Ihler, 2002). Activation of Rac is known to induce formation of lamellipodia and activation of Cdc42 induces formation of filopodia. Activated Rho is translocated from a cytoplasmic to a membrane-associated location. Rho induces formation of actin-lined invaginations in the plasma membrane. Figure 5.2 shows microspikes (Cdc42), membrane ruffling (Rac), and stress fibers (Rho) related to the activation of Rho-GTPases.

These results indicate that *Bartonella*, like other pathogenic bacteria, especially the intracellular pathogens, utilize host signaling and response mechanisms as indispensable components of the infectious process. Many other bacteria also facilitate their entry into the host cells by inducing a rearrangement of the actin cytoskeletal network (see also Chapter 6). *Shigella* entry is dependent on Cdc42, Rac, and Rho (Adam et al., 1996; Dumenil et al., 2000). *Salmonella* also uses Rac and Cdc42 in the entry process. *Salmonella* encodes a protein, the product of *sopE*, which binds to and activates Cdc42 and Rac by promoting exchange of GTP for GDP, acting as nucleotide exchange factor. A *Salmonella typhimurium* strain carrying a null mutation in *sopE* was deficient in its ability to enter cells after short infection times. A dominant negative mutant of Cdc42 prevents entry of *S. typhimurium* (Chen et al., 1996), but neither dominant negative Rac nor inhibition of Rho with C3 exoenzyme prevented membrane ruffling (Jones et al., 1993). For *Salmonella*, Cdc42 would seem to be the more important target.

Presumably activation of Rho family proteins is directly or indirectly accomplished by a *Bartonella* toxin or toxins. If it acts directly on Rho family GTPases, this hypothetical toxin could be analogous to toxins (CNF1, DNT) known in *E. coli*, *Yersinia*, and other bacteria that activate Rho proteins (for a review, see Lerm et al., 2000; Chapter 3). CNF1 readily enters cultured cells, probably by endocytosis, and is activated in the cytosol, after which it activates Rho by deamidation of glutamine 63 and has physiological effects on endothelial cells that are similar to those seen after *B. bacilliformis* infection (Vouret-Craviari et al., 1999). Epithelial cells activated with CNF1 demonstrate membrane ruffling and membrane protrusions that lead to entrapment of bacteria (or even latex particles (Falzano et al., 1993)) in large endocytic vesicles (macropinosomes). These similarities, however, arise from the fact that Rho is activated in *Bartonella* invasion and as a consequence activates downstream effector proteins, and do not necessarily imply that *Bartonella* has a toxin similar to CNF1. Indeed, other toxins, for example the *Pasteurella multocida* toxin, activate Rho indirectly by affecting upstream signaling (see Chapter 2), and *Bartonella* may operate like this.

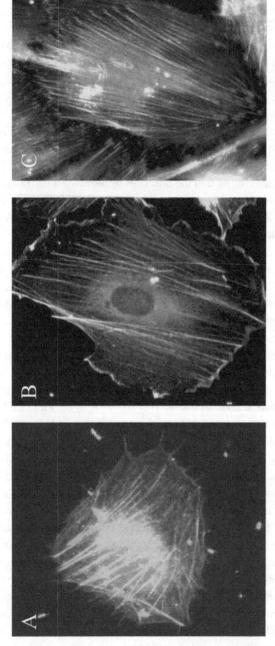

Figure 5.2. *B. bacilliformis* infected endothelial cells showing the structures associated with the activation of Rho-GTPases. A. Infected endothelial cell stained with F-actin (green) and anti-Cdc42 antibodies (red) showing microspikes related to activation of Cdc42 after 30 min of infection. B. Infected endothelial cell stained with F-actin (green) and anti-Rac antibodies (red) showing membrane ruffling and colocalization of F-actin and Rac in the membrane ruffles (yellow) related to activation of Rac after 1 h of infection. C. Infected endothelial cell stained with F-actin (green) and anti-paxillin antibodies (red) showing formation of thick stress fibers terminating in paxillin-rich focal adhesions (red) after 24 hrs of infection with the activation of Rho. (See www.cambridge.org/9780521177467 for color version.)

Both *Bartonella* and the closely related *Brucella* carry *virB* genes, coding for Type IV protein translocation systems, which are closely related to those of *Agrobacter* (Schmiederer and Anderson, 2000, Sieira et al., 2000). Inactivation of Rho, Rac, and Cdc42 with *Clostridial* toxins reduces uptake of *B. abortus* by HeLa cells, whereas CNF1 increases internalization. Dominant negative Rho, Rac, and Cdc42 inhibit uptake; the dominant positive forms promote uptake (Guzman-Verri et al., 2001).

Two Mechanisms of Entry – Conventional Endosomes and Macropinocytosis?

Caron and Hall (1998) identified two distinct pathways of macrophage phago-cytosis, an immunoglobin receptor and Cdc42/Rac-dependent pathway, and a complement receptor and Rho-dependent pathway. The existence of sepa-rate Rho-family GTPase-dependent pathways is consistent with the possibility that *B. bacilliformis* might activate multiple pathways for internalization be-cause *B. bacilliformis* activates all three Rho-GTPases. Alternatively *Bartonella* could cause the formation of a unique endocytic apparatus not directly used by the host cell, as suggested for *Shigella* (Nobes and Hall, 1995).

B. henselae has been shown to enter endothelial cells either as individual or as a few bacteria enclosed within vacuoles, or alternately in large clumps (Dehio et al., 1997; Kempf et al., 2000). Up to 24 h is required to complete internalization of the clumps, but significant numbers of individual bacteria were found to have entered within 2 h. Entry of individual *B. henselae* is not inhibited by Cytochalasin D (Kempf et al., 2001), but internalization of large clumps of bacteria by endothelial cells is prevented. These observations seem to indicate that two mechanisms of entry must be considered, entry of indi-vidual bacteria in small endosomes and entry of large clumps of bacteria by macropinocytosis. The latter mechanism might be specialized to endothelial cells.

Entry of Clumps by Macropinocytosis

A fascinating descriptive study of a unique sequence of events in the inva-sion process of endothelial cells by large clumps of *B. henselae* was reported by Dehio et al. (1997). First the leading lamella of migrating endothelial cells established contact with the bacteria and moved the bacterial aggre-gate on the cell surface by retrograde transport. Subsequently, the bacterial aggregate consisting of hundreds of bacteria was engulfed and eventually internalized in a well-defined host cell structure, called the invasome. The process was shown to be actin dependent because treatment with cytocha-lasin D prevented the engulfment and internalization of the bacteria and the whole process usually required 24 h. Membrane protrusions were found to be highly enriched with F-actin and phosphotyrosine residues with ICAM-1 redistributed to the tips of the protrusions. Stress fibers and focal adhesions were found in association with the bacteria being engulfed. A spontaneous mutant impaired in invasome formation showed an increased uptake of bac-teria into perinuclear-localizing phagosomes, presumably entering by the alternative pathway. Bacterial internalization via the invasome pathway was consistently observed with different isolates of *B. henselae*.

The invasome pathway is probably equivalent to or a variant of macropinocytosis, a process which leads to entrapment of large volumes of fluid or particles, including apoptotic cells, within intracellular vacuoles (for a review, see Swanson and Watts, 1995). Lamellipodia may fold back to enclose fluid or particles and then fuse, resulting in internalization. The membrane of the macropinosome is probably the same or similar to the plasma membrane; it is not coated and its formation is not driven by re-ceptors. The macropinosome may fuse with lysosomes in some cell types (macrophages) or may recycle back to the cell surface in others.

Macropinocytosis may be induced by CNF1 in epithelial cells, where it requires activation of Rho, Rac, and Cdc42 (Fiorentini et al., 2001). The method of internalization described by Dehio and co-workers (1997) and the involvement of Rho, Rac, and Cdc42 in internalization (Verma et al., 2000) strongly indicate that clumps of *B. bacilliformis* and *B. henselae* enter cells by a form of macropinocytosis.

Other bacteria including *Salmonella* and *Shigella* enter through a pro-cess that involves induced macropinocytosis. In particular, *Brucella*, which is closely related to *Bartonella*, has been reported to enter macrophages by a process of macropincytosis (Watarai et al., 2002).

Possible Relationship of Early Entry in Small Vesicles to Late Entry by Macropinocytosis

Bacteria that enter early might produce factors or toxins that act intracel-lularly to facilitate the entry of many more bacteria, entering as clumps by macropinocytosis. Expression of the 17-kDa antigen of the *virB* locus (Schmiederer and Anderson, 2000) was greatly stimulated in intracellular *B. heneselae* (Schmiederer et al., 2001). The authors believe that invasion and not merely attachment is required for *virB* induction, as is also the case in *Brucella suis* (Boschiroli et al., 2002). If so, this would suggest that the major activation of Rho that occurs at 12–16 h may be the consequence of intracellular induction, as well as due to the large number of bacteria present intracellularly at this time.

Bartonella and the closely related *Brucella* are each found in a perinuclear location that co-localizes with Golgi markers. Both *B. bacilliformis* and *B. abortus* associated with the Golgi are redistributed by Brefeldin A (Pizarro-Cerda et al., 1998; Verma et al., 2001). It seems likely that in *Brucella* those bacteria that enter in small vesicles induce the retrograde transport of their intracellular vesicles to the endoplasmic reticulum (Pizzarro-Cerda et al., 2000). Having reached the Golgi, the bacteria are in a position to distribute their proteins as if they were host proteins being exported and, at the same time, potentially to inject proteins or themselves into the cytoplasm of the infected cell as though through the back door. Whether this actually occurs is not known. It is, however, of considerable interest that the erythrocyte membrane, so dissimilar from the plasma membrane of nucleated cells, has similarities to Golgi membranes. Some of the membrane proteins of erythrocytes have close isoforms associated with Golgi function (Gascard and Mohandas, 2000). For example, spectrin skeleton assembly on the Golgi complex is regulated by ADP-ribosylation factor. Because *B. bacilliformis* can enter erythrocytes, it is not impossible that it may also enter the Golgi, or pass through internal membranes. Possibly tricks that evolved in an erythrocyte pathogen became useful in nucleated cells, or vice versa. It is also of some interest that erythrocytes contain several Rho family GTPases, including RhoA, whose function there is unknown but speculatively may involve modulation of spectrin and actin interactions in the erythrocyte cytoskeleton (Boukharov and Cohen, 1998).

Hewlett et al. (1994) showed that macropinosomes and early or late endosomes did not mix their contents, and that mixing was not induced by Brefeldin A. The morphology of the macropinosome was not altered by Brefeldin A, whereas conventional endosomes were altered. This suggests that the two populations of bacteria in small vesicles and macropinosomes are separate and that they may have different functions. The function of the bacteria that enter early via endosomes may be to facilitate the later entry of much larger numbers of bacteria by macropinocytosis. The function of the bacteria that enter in macropinosomes may be to provide a pool of bacteria that can readily leave the cell, but in the meantime are in a location sheltered from the immune system. The membrane of macropinosomes is derived from and is essentially the same as the cell membrane. An unusual feature of macropinosomes is that they can readily fuse with the cell membrane to release their contents to the extracellular space. This would provide

Bartonella with a sheltered location from which it could eventually exit via the macropinocytotic revolving door.

Intracellular Signaling Pathways Downstream of Rho Are Activated by *Bartonella*, Leading to Activation of Transcription Factors

Because all three Rho family GTPases are activated, some of the downstream effectors of these proteins were examined for activation as well (Verma and Ihler, 2002). The activity of PAK (serine/threonine p21–activated kinase) was increased within 1 h of addition of the bacteria, and again at 12–24 h. Stress-activated protein kinases SAPK/JNK 1 and 2 and p38 MAPK were also activated.

Activation of SAPK/JNK is known to stimulate c-Jun phosphorylation. An increased binding activity of AP-1, composed of heterodimers of Jun and Fos family proteins, as a result of c-Jun phosphorylation, was found after *B. bacilliformis* infection of endothelial cells (Verma and Ihler, 2002). AP-1 and hypoxia-inducible factor 1 cooperatively control responses to hypoxia in endothelial cells (Salnikow et al., 2002). Activation of NF-κB was not seen after infection with *B. bacilliformis*.

Activation of NF-κB by *B. henselae* did occur, however, within 10–40 min of addition of the cells and resulted in increased expression of E-selectin and ICAM-1 RNA and protein (Fuhrmann et al., 2001). Transcription of the genes for these proteins is known to require activation and translocation of NF-κB into the nucleus.

These results make it evident that *Bartonella* exert control over the host cell though activation of host programs.

Alterations of Cell Shape Associated with Activated Rho Family Proteins

Infected HUVECs, initially in a rounded shape, progressively assume an elongated or spindle-shaped morphology after 16–24 h infection. The changes in cell shape are paralleled by major actin-based changes. The unoriented F-actin network of the uninfected cells is replaced by thick F-actin bundles arranged in parallel arrays (stress fibers) oriented along the longitudinal axis of the cells. Loss of platelet endothelial cell adhesion molecule (PECAM), which localizes to cell–cell junctions, and also F-actin at cell–cell contacts becomes evident 24 h after infection. Cell–cell junctions are lost and spaces form between adjacent cells. It is likely that loss of cell–cell junctions is a mechanical consequence of the change in shape of the endothelial cells, although probably there are other contributing factors as well.

Formation of stress fibers is under the control of Rho. Fluid shear stress causes stress fiber formation and changes in focal adhesion arrangement. Platelet-activating factor alters stress fibers and causes endothelial cells to retract (Bussolino et al., 1987); sphingosine-1-phosphate released from platelets induces the formation of stress fibers in HUVECs through Rho-mediated signaling pathways (Miura et al., 2000). Thrombin-induced stress fiber formation proceeds through the thrombin receptor and heterotrimeric G proteins to Rho. Stress fibers play a role in thrombin-induced increases in permeability via intraendothelial gap formation, although permeability is also regulated by increases in cytoplasmic Ca^{+2} levels (controlled by G proteins and cyclic AMP).

The formation of stress fibers has a most profound effect on endothelial cells and the microvasculature, and plays an important role in many processes. Although stress fibers are important in angiogenesis and motility (see below), it would be premature to assume that the "intent" of the bacterial-induced stress fiber formation is angiogenesis. Loss of cell–cell contacts could be regarded as an early step in angiogenesis, in which endothelial cells sever their contacts, migrate, and proliferate in sites of new vascular synthesis. However, it could as easily be regarded as an aspect of a mechanism to induce leukocyte emigration and the loss of vascular endothelial barrier function and to establish sites of inflammation. For example, despite the fact that sphingosine-1-phosphate and thrombin each stimulate stress fiber formation, thrombin induces cell contraction and rounding whereas sphingosine-1-phosphate induces cell spreading and migration (Vouret-Craviari et al., 2002). Moreover, endostatin, an inhibitor of angiogenesis, induces tyrosine phosphorylation of FAK and paxillin and promotes formation of focal adhesions and stress fibers. We are not yet at the point where we can predict biological outcomes from information on cell signaling pathways.

Cell Motility Is Altered by Infection with *B. bacilliformis*

Macropinocytosis and cell motility are closely linked. Migrating cells display lamellipodia and filopodia at the leading edge of the cell. Substrate is attached to the leading edge of motile cells, at focal complexes (regulated by Rac and Cdc42), and at focal adhesions (regulated by Rho) where stress fibers terminate. Some proteins are common to both macropinocytosis and cell motility systems, and mutations in certain proteins inactivate both macropinocytosis and cell movement. The interconnection between lamellipodia, filopodia, and stress fibers, as well as macropinocytosis and cell motility, is strengthened by the observation that Cdc42 can activate Rac, which in turn, can activate Rho.

Endothelial cells infected with *B. bacilliformis* have a greatly reduced rate of cell migration and reduced spontaneous motility (Verma et al., 2001), and at the same time are activated for engulfment of bacteria. It appears that *B. bacilliformis*, at least temporarily, activates macropinocytosis/endocytosis pathways at the expense of cell motility.

An analogous system may exist in *Dictyostelium*. A mutant lacking *rasS* is defective in phagocytosis and fluid-phase endocytosis, and displays a three-fold greater rate of cell migration (Chubb et al., 2000). Conversely, cells overexpressing RacC displayed three-fold increased phagocytosis and three-fold reduced macropinocytosis (Seastone et al., 1998). In this system, rapid migration and macropinocytosis appear to be mutually incompatible (Sodhi et al., 2000).

Rho family proteins are integral to motility in endothelial cells. For example, sphingosine-1-phosphate promotes cell migration. Different Edg (endothelial differentiation gene) G protein-coupled receptors for sphingosine-1-phosphate regulate Rac both positively and negatively (Takuwa, 2002) and can either promote or inhibit cell migration.

SIMILARITY OF *BARTONELLA*-INDUCED CHANGES TO VASCULAR ENDOTHELIAL GROWTH FACTOR (VEGF)-INDUCED PATHWAYS

Superficially at least, many events following infection with *Bartonella* closely resemble those observed *in vivo* as a consequence of excessive VEGF stimulation. For example, mice bearing VEGF-overexpressing tumors developed hepatic pathology consisting of dilation of sinusoids, similar to those seen in peliosis hepatis associated with *B. henselae*, and marked endothelial cell proliferation with apoptosis. Resection of the tumors reversed the liver pathology, and administration of VEGF-TRAP$_{R1R2}$ – a VEGF antagonist – alleviated the pathology. Normal sinusoidal space is lined by cytoplasmic extensions of endothelial cells whose microvilli contact hepatocytes. Pathologic VEGF caused retraction of the extensions, rounding of the endothelial cells, detachment, and finally active proliferation with high rates of apoptosis.

Hypoxia is an important stimulator of angiogeneis, playing an important role in adaptation to exercise and in pathologic neovascularization, such as retinopathy of the premature newborn. Hypoxia is a potent stimulator for expression of VEGF, the key regulator of angiogenesis, including pathological angiogenesis. Two receptor tyrosine kinases, VEGFR-1 and VEGFR-2, are present on the surface of endothelial cells, and are usually upregulated under conditions of stimulated angiogenesis, so one might speculate that infection with *B. bacilliformis* might also upregulate VEGFR, but this has not yet been shown to be the case. VEGF itself can upregulate VEGFR-2 levels (Shen et al., 1998; Kremer et al., 1997). Hypoxia is a potent stimulus for post-transcriptionally increased levels of VEGFR-2 and increased functionality of the VEGF pathways (Waltenberger et al., 1996).

Binding of VEGF to VEGFR-2 induces a conformational change, dimerization, and autophosphorylation of tyrosine residues in the receptor. These phosphorylated tyrosines serve as binding sites for various target proteins, including Shc, Grb-2, c-Src, and Nck, which may themselves become phosphorylated and which activate various intracellular signaling pathways. SHP-1 and SHP-2, tyrosine phosphatases, also bind, and then deactivate VEGFR and terminate the VEGF signal.

The function of VEGFR-1 is somewhat unclear, but is probably significant because it seems to function as a negative regulator of some of the effects of VEGFR-2 and to have a role in cell migration. VEGFR-1 and -2 each have autophosphoylated tyrosine binding sites for PLCγ (Seetharam et al., 1995; Takahashi et al., 2001).

It is likely that the special tropism displayed by *Bartonella* for endothelial cells is a consequence of the existence of the VEGF pathways in endothelial cells and the major role of VEGF in endothelial cell metabolism. In fact, *Bartonella* infection does lead to VEGFR2 activation (P. Waidhet-Kouadio and G. Ihler, unpublished). It is possible that *B. bacilliformis* can directly activate VEGFR extracellularly, but it is also possible that VEGFR is activated indirectly. There is good evidence that activation of Rho results in activation upstream as well as downstream of Rho in endothelial cells; thus, activated Rho leads to phosphorylation and activation of the VEGF receptor (Gingras et al., 2000). Whether the VEGF receptor is directly activated by *B. bacilliformis* or not, it seems likely that infection with *B. bacilliformis* will activate the entire VEGF program.

The biochemical effects of VEGFR-2 mediated signal transduction include the following:

1. DNA synthesis and proliferation of cells: VEGF activation of MAP kinases ERK1/2 may occur via several pathways (Ras-Raf-MEK-ERK; PKC-Raf-MEK-ERK). JNK is also activated. Stimulation of proliferation of endothelial cells by *Bartonella* can be partially or completely explained by activation of intracellular proliferation pathways. It is interesting that VEGF is a potent mitogen for endothelial cells, but has no mitogenic

activity for other cells. Some other cell types, however, do possess VEGFR and respond to VEGF in various ways, including migration.

2. Inhibition of apoptosis: An important effect of VEGF is the enhancement of endothelial cell survival (Katoh et al., 1995; Spyridopoulos et al., 1997) by anti-apoptotic signaling. This occurs both by a PI3 kinase-dependent activation of Akt/PBK and by expression of Bcl-2 and AI, which inhibit the activation of caspases and increase synthesis of survival proteins. Inhibition of apoptosis, of course, would contribute to proliferation of endothelial cells, which was an early finding for *B. bacilliformis* and *B. henselae*.

3. Rho, Rac, and Cdc42: Infection of endothelial cells by *B. bacilliformis* is dependent on Rho and also results in activation of Rho. A similar activation is also stimulated by VEGF. Preincubation of endothelial cells with C3 exoenzyme almost entirely prevents entry of *B. bacilliformis*. When C3 exoenzyme is added at various times after the bacteria, further internalization of the bacteria is prevented. After addition of C3 exoenzyme, there is no further increase in the number of internalized bacteria and also no further increase in the percentage of cells that are infected. Thus, activation of the Rho-family portion of the VEGF pathways (and perhaps all VEGF-controlled pathways) seems to be essential to infection by *B. bacilliformis*.

4. Paxillin and focal adhesions: VEGF induces a Src-dependent tyrosine phosphorylation of FAK (Abu-Ghazaleh et al., 2001), which is required for development of new focal adhesion formation. VEGF also induces tyrosine phosphorylation of paxillin, which localizes to focal contacts. Verma et al. (2001) reported that, after *B. bacilliformis* infection, the prominent stress fibres terminated in an increased number of paxillin-containing focal contacts, located particularly at the distal ends of the longitudinal axis of infected cells, which firmly adhered the endothelial cells to the substratum.

5. Stress fibers and actin reorganization: VEGF activates p38 MAP kinase, which is on the pathway to actin reorganization because a p38 kinase inhibitor blocks actin reorganization. VEGF increases stress fiber formation. Increased formation of p38 MAP kinase and formation of actin stress fibers were observed to occur after *B. bacilliformis* infection (Verma et al., 2001; Verma and Ihler, 2002).

6. Vacuoles and lumens: In a 3-dimensional matrix, endothelial cells form vacuoles that coalesce to form lumens within 24 hrs. In this system, *B. bacilliformis*–infected endothelial cells contained numerous vesicles. But the *Bartonella*-induced vacuoles did not fuse to form lumens, and

eventually the cells died. However, in general, macropinosomes are capable of fusion with other macropinosomes (Hewlett et al., 1994). Because of the activation of Rho, Rac, and Cdc42 by *Bartonella*, the participation of these GTPases in vacuole and lumen formation was investigated. Rho was reported to play little or no role, but Cdc42 and Rac GTPases were shown to regulate intracellular vacuole and lumen formation in this system, and were targeted to the vacuolar membrane in a 3-dimensional matrix (Bayless and Davis, 2002). Possibly the intense activation of Rho by *B. bacilliformis* prevented the correct functioning of the fluid uptake and lumen-forming system.

It is possible that some minor modification of the infection protocol might have allowed vesicle fusion and lumen formation, and *in vitro* capillary formation in the infected cells; but it is perhaps more likely that *in vivo* uninfected cells, stimulated by the intense angiogenic environment, are responsible for new capillary formation and that infected cells are defective. It is, after all, asking a lot to expect *Bartonella*-infected cells to reproduce the entire complex process of angiogenesis.

In macrophages, macropinosomes eventually merge with the lysosomal compartment. However, in other cells they eventually recycle the contents to the cell surface. It seems possible that this is what is happening in endothelial cells during angiogenesis, except that the macropinosome becomes the cell surface at the lumen and the contents are recycled outside into the nascent capillary.

7. Prostaglandin I2 and NO production: VEGF stimulates endothelial production of NO and Prostaglandin I2, mediators with vascular protective effects. VEGF stimulates endothelial production of NO by increasing levels of NOS-II and NOS-III and activating them. Thus far there has not been a determination of the effect of *Bartonella* invasion on production of these and other mediators by endothelial cells, or on reactive oxygen production, but it might be reasonable to assume that these are important aspects of *Bartonella* infection.

8. Transcription factors: The physiological consequences of VEGF binding to the VEGFR1 receptor in the endothelial cell might be quite similar to *B. bacilliformis* activation of the VEGF receptor, which might occur either directly or by downstream activation of Rho-GTPases. A precedent for this is that direct injection of CNF1 into cells activated factors traditionally thought to be upstream of Rho-GTPases as well as downstream effectors7, including the VEGF receptor itself, which was phosphorylated and stimulated (Gingras et al., 2000). *B. bacilliformis*

infection does result in VEGFR2 activation (P. Waidhet-Kouadio and G. Ihler, unpublished). Consequently, all VEGF-associated pathways are likely to be activated in infected cells, whether or not Rho-GTPases lie directly in those pathways.

Other Cell Types Probably Produce VEGF, Which Stimulates Endothelial Cells in the Verruga

Direct co-cultivation of *B. henselae* with EA.hy 926 and HeLa cells resulted in formation of intracellular bacteria and also production of VEGF, whereas HUVEC, although infected, did not produce VEGF (Kempf et al., 2001). Verma et al. (2001) also reported an inability to demonstrate increased levels of VEGF from HUVEC infected with *B. bacilliformis*, although RNA levels were modestly increased. Endothelial cells do not generally express VEGF, but rather respond to VEGF. Presumably, autocrine stimulation of endothelial cells would be detrimental. This result does not necessarily mean that *Bartonella* infection of other cell types *in vivo* is necessary for VEGF production, because heat-killed *B. henselae* were effective in inducing VEGF, and LPS is capable of inducing synthesis of VEGF from some cell types (Sugishita et al., 2000). Resto-Ruiz et al. (2002) found that THP-1 macrophages infected with *B. henselae* stimulated VEGF production two-fold, whereas bacteria that had been boiled for 30 min inhibited production of VEGF. Infection also stimulated production of IL-1 beta, but not IL-8. Addition of cytochalasin D did not reduce the level of VEGF synthesis, and the absence of intracellular phagocytosed bacteria indicated that attachment was sufficient to induce VEGF and IL-1β.

The implication is that inflammation caused by various bacterial products results in local stimulation of VEGF synthesis. It seems likely that VEGF stimulation of HUVEC may play a role in activating cells for invasion.

Immune Modulation by VEGF

Long-lived persistence of the verruga is probably a key element in the *B. bacilliformis* life cycle, and permits maximal retransmission of the disease. *B. bacilliformis* itself is quite immunogenic, and survivors have life-long immunity. Consequently, the persistence of the bacteria cannot be ascribed to some stealth property of the bacterial outer membrane. Alterations in many host immunity signaling molecules are probably responsible for local immunodeficiency.

GARRET IHLER, ANITA VERMA, AND JAVIER AREVALO

One factor that is very likely to be involved is the ability of VEGF to induce immunodeficiency (Ohm and Carbone, 2001). VEGF synthesis by tumors serves to enhance the blood supply to the tumor through angiogenesis, but also the high levels of VEGF diminish the effectiveness of anti-tumor immune responses. The tumor-like properties of the verruga have been often commented upon (Arias-Stella et al., 1987) and the lesions of bacillary angiomatosis are pathologically very similar to Kaposi's sarcoma, except that Kaposi's sarcoma cannot be cured with antibiotics. A recent case of B. henselae infection was initially considered as a putative parotid gland tumor (Kempf et al., 2001), and the differential diagnosis of cat-scratch disease can include lymphoma, carcinoma, and neuroblastoma (Morbidity and Mortality Weekly Report, 2002).

Perhaps B. bacilliformis has taken advantage of the immunodeficiency-inducing properties of VEGF in the establishment of the verruga. The complete ramifications of this for understanding formation and persistence of the verruga are certainly not clear, but a suggestive mechanism can be outlined. In part, immunosuppression may be exerted on mature cells as part of an impairment in cell-mediated responses, and in part by inhibiting the maturation of cells, especially dendritic cells. Immature dendritic cells take up extracellular fluid by Rho family–dependent macropinocytosis at a rapid rate, which they constitutively process for antigens (for a review see Nobes and Marsh, 2000). Inflammatory stimuli, or bacterial components such as LPS, terminate this passive sentinel activity and cause both downregulation of macropinocytosis and dendritic cell maturation. Thus, the acquired antigens are presented only if there is other evidence of active inflammation or infection. If continued macropinocytosis prevents maturation, it may be that a B. bacilliformis toxin can cause persistence of macropinocytosis through Rho, Rac, and Cdc42. Perhaps more importantly, in dendritic cells, VEGF has been shown to inhibit NF-κB activity, which results in impaired maturation of dendritic cells (Gabrilovich et al., 1996; Oyama et al., 1998). NF-κB is also activated in dendritic cells by various bacterial stimuli, including LPS and TNFα, and for TNFα at least, VEGF has been shown to block NF-κB activation (Oyama et al., 1998). It would appear that activation as well as maturation is inhibited by VEGF. In fact, LPS augments synthesis and secretion of VEGF in myocytes (Sugishita et al., 2000), so that bacterial production of LPS may be capable of eliciting VEGF synthesis from nearby non-endothelial cells, thus activating the endothelial cells. TNFα is also an important agent for release of VEGF from various cell types.

On the other hand, LPS and TNFα can each also downregulate VEGF receptor density on endothelial cells in a soluble CD14-dependent manner

(Power et al., 2001). HUVEC also have mCD14 (Jersmann et al., 2001). Thus, it is clear that bacterial products can both stimulate synthesis of VEGF and, at the same time, reduce the number of VEGF receptors on endothelial cells. The effect of this may be to greatly increase the amount of VEGF produced, which might augment immunosuppressive effects within the verruga.

LPS initially induces synthesis of E-selectin and ICAM-1 via NF-κB activation, but persistent levels of LPS induce tolerance in HUVEC, so that there is ultimately reduced adhesion of polymorphonuclear neutrophils, as well as E-selectin, ICAM-1, and NF-κB (Lush et al., 2000).

ANTI-APOPTOSIS AND APOPTOSIS

A desirable host response to invasion of endothelial cells would presumably include death of the infected cells by apoptosis because endothelial cells die by apoptosis. Under some circumstances, infection of the HUVEC does result in cell death. For example, Verma et al. (2001) reported that when cells infected with *B. bacilliformis* were placed in an exceptionally rich medium used for *in vitro* angiogenesis and capillary formation, the infected cells died instead of performing angiogenesis. On the other hand, during infection of HUVECs, the HUVECs proliferated better than the control cells (A. Verma and G. Ihler, unpublished observations).

Several reports of enhanced proliferation were interpreted to suggest that *Bartonella* stimulates growth of endothelial cells. For example, co-cultivation of endothelial cells with *B. bacilliformis* (Garcia et al., 1992), or even with a cell-free extract from the bacteria (Garcia et al., 1990), resulted in a greater number of endothelial cells than if the cells were grown in the absence of the bacteria or the supernatant factor. This observation seemed to fit with the general expectation that *Bartonella* should stimulate endothelial cells as part of its stimulation of angiogenesis, especially because the observed effect on growth seemed to be confined to endothelial cells. This growth factor is believed to be a protein, as it has been shown to be heat sensitive and larger than 14 kDa, and can be precipitated with 45% ammonium sulfate. Thus far, however, the factor responsible for the effect remains uncharacterized. *B. henselae* stimulated proliferation and also migration of HUVEC, an effect reproduced by a trypsin-sensitive particulate fraction from the disrupted cells (Conley et al., 1994). *B. henselae* separated from HUVECs by a membrane or culture supernatants stimulated growth (Maeno et al., 1999). However, *B. henselae* did not stimulate incorporation of BrdU into endothelial cell DNA, although both mitogens VEGF and bFGF did (Kirby and Nekorchuk, 2002). Several reports also argue that *Bartonella* infection has an anti-apoptotic effect. *B. henselae* infection reduced cell death resulting from either starvation for growth factors in serum or actinomycin D (Kirby and Nekorchuk, 2002). Presumably, an absence of pro-growth stimuli is the reason that uninfected cells grown in the absence of serum die. The cells died less when infected with *Bartonella*. From this alone, it is hard to decide whether *Bartonella* provides a pro-growth or an anti-apoptotic stimulus. In fact, to a certain extent, the two are simply opposite sides of the coin.

It is presently not certain whether the stimulatory effect and the anti-apoptosis effect are one and the same, or whether there are two effects, although it is simpler to assume that they are the same. Bacterial anti-apoptosis can be achieved either by an activating signal, which would render the cells less susceptible or more resistant to apoptotic stimuli, or by direct interference with the proteins that carry out apoptosis, for example inhibition of caspases (reviewed by Hacker and Fischer, 2002). Prior to definitive caspase actions, a balancing of pro- and anti-apoptotic stimuli determines the fate of the cell. The effect of increased NF-κB activity is well known to provide protection against apoptosis by mechanisms that include increased synthesis of anti-apoptotic proteins. In addition, the activation of Rho family proteins, and the resulting changes in cell morphology/function that occur, would be expected to have an anti-apoptotic effect. For example sphingosine-1-phosphate acting through heterotrimeric G protein-linked receptors both stimulates cell proliferation and provides protection from apoptosis (Hisano et al., 1999). Thus *Bartonella* would automatically promote anti-apoptosis by activating Rho-dependent and other pathways.

Because bacterial products (e.g., LPS) have apoptosis-inducing effects, to a certain extent *Bartonella* may be fighting itself. Why not simply leave out both the protecting and inducing molecules? Possibly, molecules that induce apoptosis are intended for non-endothelial cells (e.g. lymphocytes) and protection is extended only to infected endothelial cells.

OTHER ROLES OF RHO-GTPASES RELATING TO IMMUNITY

Rho-GTPases are important to a large extent because actin is important. Because of the central role of actin, numerous aspects of endothelial cell function depend on Rho-family GTPases. One of the most important functions of endothelial cells is facilitation of extravasation of leukocytes by arresting their progress through the expression (upregulated by inflammatory cytokines) and clustering of leukocyte binding proteins and by reduced adherence to neighboring endothelial cells. For example, clustering of monocyte receptors E-selectin, ICAM-1, and VCAM-1 and their linkage to the actin cytoskeleton are modulated by Rho (but not Rac or Cdc42) (Wojciak-Stothard et al., 1999), and so inhibition of Rho results in impaired adhesion of monocytes. Bacteria that inactivate Rho may thus readily achieve impairment of crucial Rho-dependent functions such as monocyte adherence. Bacteria like *Bartonella*, which activate Rho, presumably do so to induce the endothelial cells to perform functions not present in quiescent cells. Introduction of constitutively activated Rho into quiescent HUVECs does not increase adhesion of monocytes, unless they had been stimulated with TNFα (Wojciak-Stothard et al., 1999). Potentially, it could be possible for bacteria such as *Bartonella* to activate Rho without enhancing leukocyte adhesion, but in fact *B. henselae* markedly upregulates E-selectin (at early times) and ICAM-1 (at later times) in an NF-κB dependent manner. This stimulation can be effected by outer membrane proteins or by sonicated extracts and does not require live bacteria or bacterial infection, because the nonpiliated strain, unable to invade endothelial cells, induced ICAM-1 (Fuhrmann et al., 2001; Maeno et al., 2002;). The presence of these adhesive endothelial proteins markedly enhances rolling and adhesion of leukocytes.

IL-8, an angiogenic and neutrophil-recruitment factor, acts through two IL-8 receptors that are present on endothelial cells. Through IL-8 receptor type-1, IL-8 activates Rho leading to actin polymerization and stress fiber formation. Rac is activated through IL-8 receptor type-2 activation and leads to cell retraction and gap formation between adjacent cells (Schraufstatter et al., 2001). The importance of IL-8 to endothelial cell function is emphasized by the observation that IL-8 is stored in Weibel-Palade bodies in some endothelial cells (but not HUVEC) and is released by inflammatory stimuli such as histamine or thrombin (Utgaard et al., 1998).

FUTURE DIRECTIONS

The lesions of *B. bacilliformis* and *B. henselae* are long lasting and must involve manipulation and modulation of the host immune response. Because these infections are ultimately benign and curable and yet present some of the complex biological features of neoplastic growth, they may provide a useful model system in humans for the testing and further development of therapies and drugs that target neoplastic angiogenesis. Increased knowledge about the interplay of host and bacterial signaling systems may help to unravel both the remaining mysteries of angiogenesis and the apparently close relationship between angiogenesis and tumor biology.

ACKNOWLEDGEMENT

This work was supported by the Tom and Jean McMullin Chair in Genetics. We thank Dr. Doug Struck for his review of this manuscript.

REFERENCES

Abu-Ghazaleh R, Kabir J, Jia H, Lobo M, and Zachary I (2001). Src mediates stimulation by vascular endothelial growth factor of the phosphorylation of focal adhesion kinase at tyrosine 861, and migration and anti-apoptosis in endothelial cells. *Biochem. J.*, **360**, 255–264.

Adam T, Giry M, Boquet P, and Sansonetti P (1996). Rho-dependent membrane folding causes *Shigella* entry into epithelial cells. *EMBO J.*, **15**, 3315–3321.

Arias-Stella J, Lieberman P H, Erlandson R A, and Arias-Stella J Jr (1986). Histology, immunohistochemistry, and ultrastructure of the verruga in Carrion's disease. *Am. J. Surg. Pathol.*, **10**, 595–610.

Arias-Stella J, Lieberman P H, Garcia-Caceres U, Erlandson R A, Kruger H, and Arias-Stella J Jr (1987). Verruga peruana mimicking malignant neoplasms. *Am. J. Dermatopath.*, **9**, 279–291.

Bayless K J and Davis G E (2002). The Cdc42 and Rac1 GTPases are required for capillary lumen formation in three-dimensional extracellular matrices. *J. Cell Sci.*, **115**, 1123–1136.

Benson L A, Kar S, McLaughlin G, and Ihler G M (1986). Entry of *Bartonella bacilliformis* into erythrocytes. *Infect. Immun.*, **54**, 347–353.

Boschiroli M L, Ouahrani-Bettache S, Foulongne V, Michaux-Charachon S, Bourg G, Allardet-Servent A, Cazevieille C, Liautard J P, Ramuz M, and O'Callaghan D (2002). The *Brucella suis virB* operon is induced intracellularly in macrophages. *Proc. Natl. Acad. Sci. USA*, **99**, 1544–1549.

Boukharov A A and Cohen C M (1998). Guanine nucleotide-dependent translocation of RhoA from cytosol to high affinity membrane binding sites in human erythrocytes. *Biochem. J.*, **330**, 1391–1398.

Breitschwerdt E B and Kordick D L (2000). *Bartonella* infection in animals: Carriership, reservoir potential, pathogenicity, and zoonotic potential for human infection. *Clin. Microbiol. Rev.*, **13**, 428–438.

Buckles E L and McGinnis Hill E (2000). Interaction of *Bartonella bacilliformis* with human erythrocyte membrane proteins. *Microb. Pathogenesis.*, **29**, 165–174.

Burgess A W, Paquet J Y, Letesson J J, and Anderson B E (2000). Isolation, sequencing and expression of *Bartonella henselae* omp43 and predicted membrane topology of the deduced protein. *Microb. Pathogenesis*, **29**, 73–80.

Bussolino F, Camussi G, Aglietta M, Braquet P, Bosia A, Pescarmona G, Sanavio F, D'Urso N, and Marchisio P C (1987). Human endothelial cells are target for platelet-activating factor. I. Platelet-activating factor induces changes in cytoskeleton structures. *J. Immunol.*, **139**, 2439–2446.

Caron E and Hall A (1998). Identification of two distinct mechanisms of phagocytosis controlled by different Rho GTPases. *Science*, **282**, 1717–1721.

Chen L M, Hobbie S and Galan J E (1996). Requirement of CDC42 for *Salmonella*-induced cytoskeletal and nuclear responses. *Science*, **274**, 2115–2118.

Chubb J R, Wilkins A, Thomas G M, and Insall R H (2000). The *Dictyostelium RasS* protein is required for macropinocytosis, phagocytosis and the control of cell movement. *J. Cell Sci.*, **113**, 709–719.

Closse C, Dachary-Prigent J, and Boisseau M R (1999). Phosphatidylserine-related adhesion of human erythrocytes to vascular endothelium. *Brit. J. Haematol.*, **107**, 300–302.

Conley T, Slater L, and Hamilton K (1994). *Rochalimaea* species stimulate human endothelial cell proliferation and migration *in vitro*. *J. Lab. Clin. Med.*, **124**, 521–528.

Dehio C (1999). Interactions of *Bartonella henselae* with vascular endothelial cells. *Curr. Opin. Microbiol.*, **2**, 78–82.

Dehio C (2001). *Bartonella* interactions with endothelial cells and erythrocytes. *Trends Microbiol.*, **9**, 279–285.

Dehio C, Meyer M, Berger J, Schwarz H, and Lanz C (1997). Interaction of *Bartonella henselae* with endothelial cells results in bacterial aggregation on the cell surface and the subsequent engulfment and internalisation of the bacterial aggregate by a unique structure, the invasome. *J. Cell Sci.*, **110**, 2141–2154.

Derrick S C and Ihler G M (2001). Deformin, a substance found in *Bartonella bacilliformis* culture supernatants, is a small, hydrophobic molecule with an affinity for albumin. *Blood Cell. Mol. Dis.*, **27**, 1013–1019.

Dumenil G, Sansonetti P, and Tran Van Nhieu G (2000). Src tyrosine kinase activity down-regulates Rho-dependent responses during *Shigella* entry into epithelial cells and stress fibre formation. *J. Cell Sci.*, **113**, 71–80.

Earhart K C and Power M H (2000). Images in clinical medicine. *Bartonella neuroretinitis*. *N. Engl. J. Med.*, **343**, 1459.

Falzano L, Fiorentini C, Donelli G, Michel E, Kocks C, Cossart P, Cabanie L, Oswald E, and Boquet P (1993). Induction of phagocytic behaviour in human epithelial cells by *Escherichia coli* cytotoxic necrotizing factor type 1. *Mol. Microbiol.*, **9**, 1247–1254.

Fiorentini C, Falzano L, Fabbri A, Stringaro A, Logozzi M, Travaglione S, Contamin S, Arancia G, Malorni W, and Fais S (2001). Activation of rho GTPases by cytotoxic necrotizing factor 1 induces macropinocytosis and scavenging activity in epithelial cells. *Mol. Biol. Cell.*, **12**, 2061–2073.

Fuhrmann O, Arvand M, Gohler A, Schmid M, Krull M, Hippenstiel S, Seybold J, Dehio C, and Suttorp N (2001). *Bartonella henselae* induces NF-?B-dependent upregulation of adhesion molecules in cultured human endothelial cells: Possible role of outer membrane proteins as pathogenic factors. *Infect. Immun.*, **69**, 5088–5097.

Gabrilovich D I, Chen H L, Girgis K R, Cunningham H T, Meny G M, Nadaf S, Kavanaugh D, and Carbone D P (1996). Production of vascular endothelial growth factor by human tumors inhibits the functional maturation of dendritic cells. *Nat. Med.*, **2**, 1096–1103.

Garcia F U, Wojta J, Broadley K N, Davidson J M, and Hoover R L (1990). *Bartonella bacilliformis* stimulates endothelial cells *in vitro* and is angiogenic *in vivo*. *Am. J. Pathol.*, **136**, 1125–1135.

Garcia F U, Wojta J, and Hoover R L (1992). Interactions between live *Bartonella bacilliformis* and endothelial cells. *J. Infect. Dis.*, **165**, 1138–1141.

Gascard P and Mohandas N (2000). New insights into functions of erythroid proteins in nonerythroid cells. *Curr. Opin. Hematol.*, **7**, 123–129.

Gingras D, Lamy S, and Beliveau R (2000). Tyrosine phosphorylation of the vascular endothelial-growth-factor receptor-2 (VEGFR-2) is modulated by Rho proteins. *Biochem. J.*, **348**, 273–280.

Guptill L, Wu C C, Glickman L, Turek J, Slater L, and HogenEsch H (2000). Extracellular *Bartonella henselae* and artifactual intraerythrocytic pseudoinclusions in experimentally infected cats. *Vet. Microbiol.*, **76**, 283–290.

Guzman-Verri C, Chaves-Olarte E, von Eichel-Streiber C, Lopez-Goni I, Thelestam M, Arvidson S, Gorvel J P, and Moreno E (2001). GTPases of the Rho subfamily are required for *Brucella abortus* internalization in nonprofessional phagocytes: Direct activation of Cdc42. *J. Biol. Chem.*, **276**, 44435–44443.

Hacker G and Fischer S F (2002). Bacterial anti-apoptotic activities. *FEMS Microbiol. Lett.*, **211**, 1–6.

Harsch I, Schahin S, Schmelzer A, Hahn E, and Konturek P (2002). Cat-scratch disease in an immunocompromised host. *Med. Sci. Monit.*, **8**, CS26–29.

Hewlett L J, Prescott A R, and Watts C (1994). The coated pit and macropinocytic pathways serve distinct endosome populations. *J. Cell Biol.*, **124**, 689–703.

Hill E M, Raji A, Valenzuela M S, Garcia F, and Hoover R (1992). Adhesion to and invasion of cultured human cells by *Bartonella bacilliformis*. *Infect Immun.*, **60**, 4051–4058.

Hisano N, Yatomi Y, Satoh K, Akimoto S, Mitsumata M, Fujino M A, and Ozaki Y (1999). Induction and suppression of endothelial cell apoptosis by sphingolipids: A possible *in vitro* model for cell-cell interactions between platelets and endothelial cells. *Blood*, **93**, 4293–4299.

Houpikian P and Raoult D (2001). Molecular phylogeny of the genus *Bartonella*: What is the current knowledge? *FEMS Microbiol. Lett.*, **200**, 1–7.

Ihler G M (1996). *Bartonella bacilliformis*: Dangerous pathogen slowly emerging from deep background. *FEMS Microbiol. Lett.*, **144**, 1–11.

Iwaki-Egawa S and Ihler G M (1997). Comparison of the abilities of proteins from *Bartonella bacilliformis* and *Bartonella henselae* to deform red cell membranes and to bind to red cell ghost proteins. *FEMS Microbiol. Lett.*, 157, 207–217.

Jersmann H P A, Hii C S T, Hodge G L, and Ferrante A (2001). Synthesis and surface expression of CD14 by human endothelial cells. *Infect. Immun.*, 69, 479–485.

Jones B D, Paterson H F, Hall A, and Falkow S (1993). *Salmonella typhimurium* induces membrane ruffling by a growth factor-receptor-independent mechanism. *Proc. Natl. Acad. Sci. USA*, 90, 10390–10394.

Karem K L (2000). Immune aspects of *Bartonella*. *Crit. Rev. Microbiol.*, 26, 133–145.

Karem K L, Paddock C D, and Regnery R L (2000). *Bartonella henselae, B. quintana,* and *B. bacilliformis*: Historical pathogens of emerging significance. *Microbes. Infect.*, 2, 1193–1205.

Katoh O, Tauchi H, Kawaishi K, Kimura A, and Satow Y (1995). Expression of the vascular endothelial growth factor (VEGF) receptor gene, KDR, in hematopoietic cells and inhibitory effect of VEGF on apoptotic cell death caused by ionizing radiation. *Cancer Res.*, 55, 5687–5692.

Kempf V A J, Petzold H, and Autenrieth I B (2001). Cat scratch disease due to *Bartonella henselae* infection mimicking parotid malignancy. *Eur. J. Clin. Microbiol.*, 20, 732–733.

Kempf V A J, Schaller M, Behrendt S, Volkmann B, Aepfelbacher M, Cakman I, and Autenrieth I B (2000). Interaction of *Bartonella henselae* with endothelial cells results in rapid bacterial rRNA synthesis and replication. *Cell. Microbiol.*, 2, 431–441.

Kempf V A J, Volkmann B, Schaller M, Sander C A, Alitalo K, Riess T, and Autenrieth I B (2001). Evidence of a leading role for VEGF in *Bartonella henselae*-induced endothelial cell proliferations. *Cell. Microbiol.*, 3, 623–632.

Kirby J E and Nekorchuk D M (2002). *Bartonella*-associated endothelial proliferation depends on inhibition of apoptosis. *Proc. Natl. Acad. Sci. USA*, 99, 4656–4661.

Kordick D L and Breitschwerdt E B (1995). Intraerythrocytic presence of *Bartonella henselae*. *J. Clin. Microbiol.*, 33, 1655–1656.

Kostianovsky M, Lamy Y, and Greco M A (1992). Immunohistochemical and electron microscopic profiles of cutaneous Kaposi's sarcoma and bacillary angiomatosis. *Ultrastruct. Pathol.*, 16, 629–640.

Kremer C, Breier G, Risau W, and Plate K H (1997). Up-regulation of flk-1/vascular endothelial growth factor receptor 2 by its ligand in a cerebral slice culture system. *Cancer Res.*, 57, 3852–3859.

LeBoit P E, Berger T G, Egbert B M, Beckstead J H, Yen T S, and Stoler M H (1989). Bacillary angiomatosis. The histopathology and differential diagnosis

of a pseudoneoplastic infection in patients with human immunodeficiency
virus disease. *Am. J. Surg. Pathol.*, **13**, 909–920.

Lee J and Lynde C (2001). Pyogenic granuloma: Pyogenic again? Association
between pyogenic granuloma and *Bartonella*. *J. Cutan. Med. Surg.*, **5**, 467–
470.

Lerm M, Schmidt G, and Aktories K (2000). Bacterial protein toxins targeting rho
GTPases. *FEMS Microbiol. Lett.*, **188**, 1–6.

Lush C W, Cepinskas G, and Kvietys P R (2000). LPS tolerance in human endothelial cells: Reduced PMN adhesion, E-selectin expression, and NF-κB
mobilization. *Am. J. Physiol. Heart. C.*, **278**, H853–861.

Maeno N, Oda H, Yoshiie K, Wahid M R, Fujimura T, and Matayoshi S (1999).
Live *Bartonella henselae* enhances endothelial cell proliferation without direct
contact. *Microb. Pathogenesis.*, **27**, 419–427.

Maeno N, Yoshiie K, Matayoshi S, Fujimura T, Mao S, Wahid M R, and Oda H
(2002). A heat-stable component of *Bartonella henselae* upregulates intercellular adhesion molecule-1 expression on vascular endothelial cells. *Scand. J.
Immunol.*, **55**, 366–372.

Maguina C and Gotuzzo E (2000). Bartonellosis – new and old. *Infect. Dis. Clin.
N. Am.*, **14**, 1–22.

Massari P, Henneke P, Ho Y, Latz E, Golenbock D T, and Wetzler L M (2002).
Cutting edge: Immune stimulation by neisserial porins is toll-like receptor
2 and MyD88 dependent. *J. Immunol.*, **168**, 1533–1537.

Minnick M F and Anderson B E (2000). *Bartonella* interactions with host cells.
Sub-Cell. Biochem., **33**, 97–123.

Miura Y, Yatomi Y, Rile G, Ohmori T, Satoh K, and Ozaki Y (2000). Rho-mediated
phosphorylation of focal adhesion kinase and myosin light chain in human endothelial cells stimulated with sphingosine 1-phosphate, a bioactive
lysophospholipid released from activated platelets. *J. Biochem.-Tokyo*, **127**,
909–914.

Morbidity and Mortality Weekly Report. (2002). Cat-scratch disease in children –
Texas, September 2000–August 2001. *MMWR Morb. Mortal Wkly. Rep.*, **51**,
212–214.

Muller A, Rassow J, Grimm J, Machuy N, Meyer T F, and Rudel T (2002). VDAC
and the bacterial porin PorB of *Neisseria gonorrhoeae* share mitochondrial
import pathways. *EMBO J.*, **21**, 1916–1929.

Nobes C D and Hall A (1995). Rho, rac, and cdc42 GTPases regulate the assembly of multimolecular focal complexes associated with actin stress fibers,
lamellipodia, and filopodia. *Cell*, **81**, 53–62.

Nobes C and Marsh M (2000). Dendritic cells: New roles for Cdc42 and Rac in
antigen uptake? *Curr. Biol.*, **10**, R739–41.

Ohm J E and Carbone D P (2001). VEGF as a mediator of tumor-associated immunodeficiency. *Immunol. Res.*, **23**, 263–272.

Oyama T, Ran S, Ishida T, Nadaf S, Kerr L, Carbone D P, and Gabrilovich D I (1998). Vascular endothelial growth factor affects dendritic cell maturation through the inhibition of nuclear factor-kappa B activation in hemopoietic progenitor cells. *J. Immunol.*, **160**, 1224–1232.

Pizarro-Cerda J, Meresse S, Parton R G, vanderGoot G, Sola-Landa A, Lopez-Goni I, Moreno E, and Gorvel J P (1998). *Brucella abortus* transits through the autophagic pathway and replicates in the endoplasmic reticulum of non-professional phagocytes. *Infect. Immun.*, **66**, 5711–5724.

Pizarro-Cerda J, Moreno E, and Gorvel J P (2000). Invasion and intracellular trafficking of *Brucella abortus* in nonphagocytic cells. *Microbes Infect.*, **2**, 829–835.

Power C, Wang J H, Sookhai S, Street J T, and Redmond H P (2001). Bacterial wall products induce downregulation of vascular endothelial growth factor receptors on endothelial cells via a CD14-dependent mechanism: Implications for surgical wound healing. *J. Surg. Res.*, **101**, 138–145.

Resto-Ruiz S I, Schmiederer M, Sweger D, Newton C, Klein T W, Friedman H, and Anderson B E (2002). Induction of a potential paracrine angiogenic loop between human THP-1 macrophages and human microvascular endothelial cells during *Bartonella henselae* infection. *Infect. Immun.*, **70**, 4564–4570.

Rolain J M, LaScola B, Liang Z, Davoust B, and Raoult D (2001). Immunofluorescent detection of intraerythrocytic *Bartonella henselae* in naturally infected cats. *J. Clin. Microbiol.*, **39**, 2978–2980.

Salnikow K, Kluz T, Costa M, Piquemal D, Demidenko Z N, Xie K, and Blagosklonny M V (2002). The regulation of hypoxic genes by calcium involves c-Jun/AP-1, which cooperates with hypoxia-inducible factor 1 in response to hypoxia. *Mol. Cell. Biol.*, **22**, 1734–1741.

Schmiederer M, Arcenas R, Widen R, Valkov N, and Anderson B (2001). Intracellular induction of the *Bartonella henselae virB* operon by human endothelial cells. *Infect. Immun.*, **69**, 6495–6502.

Schmiederer M and Anderson B (2000). Cloning, sequencing, and expression of three *Bartonella henselae* genes homologous to the *Agrobacterium tumefaciens* VirB region. *DNA Cell Biol.*, **19**, 141–147.

Schraufstatter I U, Chung J, and Burger M (2001). IL-8 activates endothelial cell CXCR1 and CXCR2 through Rho and Rac signaling pathways. *Am. J. Physiol.-Lung C.*, **280**, L1094–1103.

Schulein R, Seubert A, Gille C, Lanz C, Hansmann Y, Piemont Y, and Dehio C (2001). Invasion and persistent intracellular colonization of erythrocytes. A unique parasitic strategy of the emerging pathogen *Bartonella. J. Exp. Med.*, **193**, 1077–1086.

Seastone D J, Lee E, Bush J, Knecht D, and Cardelli J (1998). Overexpression of a novel rho family GTPase, RacC, induces unusual actin-based structures and positively affects phagocytosis in *Dictyostelium discoideum. Mol. Biol. Cell*, **9**, 2891–2904.

Seetharam L, Gotoh N, Maru Y, Neufeld G, Yamaguchi S, and Shibuya M (1995). A unique signal transduction from FLT tyrosine kinase, a receptor for vascular endothelial growth factor VEGF. *Oncogene*, **10**, 135–147.

Shen B Q, Lee D Y, Gerber H P, Keyt B A, Ferrara N, and Zioncheck T F (1998). Homologous up-regulation of KDR/Flk-1 receptor expression by vascular endothelial growth factor *in vitro. J. Biol. Chem.*, **273**, 29979–29985.

Sieira R, Comerci D J, Sanchez D O, and Ugalde R A (2000). A homologue of an operon required for DNA transfer in *Agrobacterium* is required in *Brucella abortus* for virulence and intracellular multiplication. *J. Bacteriol.*, **182**, 4849–4855.

Sodhi A, Montaner S, Patel V, Zohar M, Bais C, Mesri E A, and Gutkind J S (2000). The Kaposi's sarcoma-associated herpes virus G protein-coupled receptor up-regulates vascular endothelial growth factor expression and secretion through mitogen-activated protein kinase and p38 pathways acting on hypoxia-inducible factor 1a. *Cancer Res.*, **60**, 4873–4880.

Spyridopoulos I, Brogi E, Kearney M, Sullivan A B, Cetrulo C, Isner J M, and Losordo D W (1997). Vascular endothelial growth factor inhibits endothelial cell apoptosis induced by tumor necrosis factor-alpha: Balance between growth and death signals. *J. Mol. Cell. Cardiol.*, **29**, 1321–1330.

Sugishita Y, Shimizu T, Yao A, Kinugawa K, Nojiri T, Harada K, Matsui H, Nagai R, and Takahashi T (2000). Lipopolysaccharide augments expression and secretion of vascular endothelial growth factor in rat ventricular myocytes. *Biochem. Biophys. Res. Commun.*, **268**, 657–662.

Swanson J A and Watts C (1995). Macropinocytosis. *Trends Cell Biol.*, **5**, 424–428.

Takahashi T, Yamaguchi S, Chida K, and Shibuya M (2001). A single autophosphorylation site on KDR/Flk-1 is essential for VEGF-A-dependent activation of PLC-gamma and DNA synthesis in vascular endothelial cells. *EMBO J.*, **20**, 2768–2778.

Takuwa Y (2002). Subtype-specific differential regulation of Rho family G proteins and cell migration by the Edg family sphingosine-1-phosphate receptors. *Biochim. Biophys. Acta*, **1582**, 112–120.

Tsukahara M, Iino H, Ishida C, Murakami K, Tsuneoka H, and Uchida M (2001). *Bartonella henselae* bacteraemia in patients with cat scratch disease. *Eur. J. Pediatr.*, **160**, 316.

Utgaard J O, Jahnsen F L, Bakka A, Brandtzaeg P, and Haraldsen G (1998). Rapid secretion of prestored interleukin 8 from Weibel-Palade bodies of microvascular endothelial cells. *J. Exp. Med.*, **188**, 1751–1756.

Verma A, Davis G E, and Ihler G M (2001). Formation of stress fibres in human endothelial cells infected with *Bartonella bacilliformis* is associated with altered morphology, impaired migration and defects in cell morphogenesis. *Cell. Microbiol.*, **3**, 169–180.

Verma A, Davis G E, and Ihler G M (2000). Infection of human endothelial cells with *Bartonella bacilliformis* is dependent on Rho and results in activation of Rho. *Infect. Immun.*, **68**, 5960–5969.

Verma A and Ihler G M (2002). Activation of Rac, Cdc42 and other downstream signalling molecules by *Bartonella bacilliformis* during entry into human endothelial cells. *Cell. Microbiol.*, **4**, 557–569.

Vouret-Craviari V, Bourcier C, Boulter E, and van Obberghen-Schilling E (2002). Distinct signals via Rho GTPases and Src drive shape changes by thrombin and sphingosine-1-phosphate in endothelial cells. *J. Cell Sci.*, **115**, 2475–2484.

Vouret-Craviari V, Grall D, Flatau G, Pouyssegur J, Boquet P, and Van Obberghen-Schilling E (1999). Effects of cytotoxic necrotizing factor 1 and lethal toxin on actin cytoskeleton and VE-cadherin localization in human endothelial cell monolayers. *Infect. Immun.*, **67**, 3002–3008.

Waltenberger J, Mayr U, Pentz S, and Hombach V (1996). Functional upregulation of the vascular endothelial growth factor receptor KDR by hypoxia. *Circulation*, **94**, 1647–1654.

Watarai M, Makino S, Fujii Y, Okamoto K, and Shirahata T (2002). Modulation of *Brucella*-induced macropinocytosis by lipid rafts mediates intracellular replication. *Cell. Microbiol.*, **4**, 341–355.

Wojciak-Stothard B, Williams L, and Ridley A J (1999). Monocyte adhesion and spreading on human endothelial cells is dependent on Rho-regulated receptor clustering. *J. Cell Biol.*, **145**, 1293–1307.

CHAPTER 6

Type III–delivered toxins that target signalling pathways

Luís J Mota and Guy R Cornelis

Upon infection, pathogenic bacteria must evade the immune defence of their host in order to multiply. To this end, many bacteria secrete toxins as part of their virulence mechanism. In a classical view, toxins are molecules that cause intoxication upon their release by bacteria into the body fluids. However, in the last 10 years a different class of bacterial toxin has been recognised. These molecules are not simply secreted by the bacterium, but instead they are delivered directly from the bacterial cytoplasm into the cytoplasm of the eukaryotic cell by specialised secretion machines present exclusively in Gram-negative bacteria. These are the so-called type III or type IV secretion systems, depending on whether they use a structure resembling the flagella or conjugative pili, respectively. In this chapter, we will describe the mode of action of toxins delivered by type III secretion systems (TTSSs). These molecules, currently known as type III effectors, have been shown to act on different host signalling pathways controlling a number of responses, and in some cases interfere with cell growth.

TYPE III SECRETION SYSTEMS

TTSSs are present not only in bacteria that are pathogenic for animals but also in bacteria pathogenic for plants or even in symbionts for plants and insects (Cornelis and Van Gijsegem, 2000). We will restrict our analysis to the action of type III effectors of animal pathogens. Among these, type III effectors have been identified in *Yersinia* spp., in *Salmonella* spp., in *Shigella* spp., in enteropathogenic and enterohaemorrhagic *Escherichia coli*, in *Pseudomonas aeruginosa*, and more recently, in *Burkholderia pseudomallei* (Stevens et al., 2003). For comprehensive reviews covering the different aspects of

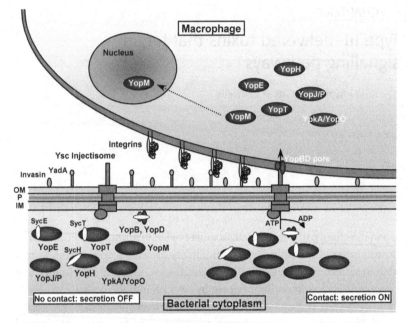

Figure 6.1. The *Yersinia* Ysc-Yop model of type III secretion. When *Yersinia* is placed at the temperature of its host, a needle-like structure called the Ysc injectisome is assembled and a stock of Yop proteins is produced. Some Yops are kept in the cytoplasm bound to their specific Syc chaperone. In the absence of contact with a eukaryotic cell, the secretion channel is closed (secretion OFF). Upon contact with a eukaryotic cell, the bacterial adhesins Invasin and YadA interact with integrins at the surface of the eukaryotic cell, which docks the bacterium at the cell's surface and promotes opening of the secretion channel (secretion ON). YopB and YopD form a pore in the target cell plasma membrane; and the Yop effectors are delivered into the eukaryotic cell cytosol through this pore. Among the effectors, YopM is further translocated into the cell nucleus. The Yop proteins are not drawn to scale. OM, outer membrane; P, peptidoglycan; IM, inner membrane. Adapted from Cornelis (2002).

TTSSs and a more complete list of references see Cornelis and Van Gijsegem (2000) and Buttner and Bonas (2002).

The physiological function of TTSSs is to deliver bacterial proteins into eukaryotic cells. This is accomplished by a complex secretion system. The proteins not only are secreted across the two bacterial membranes but are also translocated across the eukaryotic cell membrane. The Ysc-Yop TTSS of *Yersinia* is one of the best studied and provides a good model to understand how type III secretion works (Figure 6.1) (Cornelis, 2002). The system consists of secreted proteins, called Yops, and their dedicated type III secretion apparatus, a needle-like complex called the Ysc injectisome. The length of the needle is controlled by the YscP protein, which was recently shown to act as a molecular ruler (Journet et al., 2003). The Yop proteins include intracellular "effectors" (YopE, YopH, YopM, YpkA/YopO, YopJ/P, YopT) and "translocators" (YopB, YopD, LcrV), which form a pore in the host cell plasma membrane and are needed to deliver the effectors into the cytosol of eukaryotic target cells. The system also secretes proteins that seem to have an exclusive regulatory role in the bacteria (YopN, YopQ, YscM/LcrQ), components of the Ysc injectisome itself (YscP, YscF), and one protein with unknown function (YopR). Secretion of some Yops requires the assistance, in the bacterial cytosol, of small individual type III chaperones, called the Syc proteins, which bind specifically to their cognate Yop. One of the hallmarks of type III secretion is that its substrates have no classical cleaved NH_2-terminal signal sequence, but Yops are nevertheless recognised by the NH_2-terminus. Furthermore, type III secretion is a contact-dependent phenomenon, and physiological secretion of Yops is triggered by intimate contact between an invading bacterium and a target cell. In general, these basic principles are shared by the other TTSSs.

THE ACTION OF TYPE III EFFECTORS ON HOST SIGNALLING PATHWAYS

The mode of action and biochemical activities of type III effectors identified to date are described below, in the context of the different bacterial pathogenesis mechanisms.

Type III Effectors Acting on Small GTP-binding Proteins

Small GTP-binding proteins act as molecular switches to regulate many essential cellular processes (Takai et al., 2001). Therefore, they are ideal targets for bacterial toxins (Boquet, 2000; see also Chapter 3), and type III effectors are no exception. Small GTP-binding proteins constitute a superfamily structurally classified into distinct groups: the Ras, Rho, Rab, Sar1/Arf, and Ran families. Through gene expression regulation, Ras proteins control cell proliferation, differentiation, morphology, and apoptosis. The Rho proteins are master regulators of cytoskeleton dynamics and also control gene expression. The Rab and Sar1/Arf families control vesicle trafficking and Ran proteins regulate nucleocytoplasmic transport and microtubule organisation. Small GTP-binding proteins cycle between an inactive GDP-bound state and an activated GTP-bound state. Activation occurs by GDP and GTP exchange, a process promoted by guanine nucleotide exchange factors (GEFs). Inactivation occurs by GTP hydrolysis through the action of their intrinsic GTPase activity, a process facilitated by GTPase activating proteins (GAPs). When activated, the small GTP binding proteins interact with a wide variety of effector proteins to mediate downstream signalling. Small GTP-binding proteins of the Ras, Rho, and Rab families have sequences at their carboxyl termini that undergo post-translational modifications with lipid, such as farnesyl or geranygeranyl. This lipid modification is required for their binding to membranes and activation of downstream effectors.

In the case of type III effectors, a recurring theme is modulation of cytoskeleton dynamics, either to promote bacterial entry into host cells or to prevent uptake by phagocytic cells. It is therefore not surprising that small GTP-binding proteins of the Rho family (RhoGTPases) are major targets of type III effectors. In addition, type III effectors have been found to act on proteins from the Ras and Rab family.

Type III Effectors That Modulate RhoGTPase Signalling and Promote Bacterial Entry

The ability to enter cells that are normally non-phagocytic, such as those that line the intestinal epithelium, is an essential step in the pathogenesis of *Salmonella* and *Shigella*. In each case the entry of these intracellular pathogens is promoted by a set of type III effectors that either directly or indirectly modulate RhoGTPases. The most extensively characterised RhoGTPases are Cdc42, Rac1, and RhoA (Hall, 1998). Cdc42 induces the formation of actin-rich finger-like protrusions called filopodia, whereas Rac1 determines the formation of pseudopodial structures called lamellipodia and of membrane ruffles. RhoA induces the formation of actin stress fibres and focal adhesions.

The Concerted Action of *Salmonella* Effectors on RhoGTPases

Salmonella enterica encodes two TTSSs located at discrete regions of its chromosome (pathogenicity islands 1 and 2, named SPI-1 and SPI-2) that are essential for pathogenicity (reviewed by Galán, 2001). While the SPI-1 encoded TTSS is required for the initial interaction of *Salmonella* with the intestinal epithelial cells, SPI-2 is required for systemic infection.

The interaction of *Salmonella* with cultured epithelial cells triggers signal transduction pathways that lead to a variety of cellular responses (Galán, 2001). One of these responses is characterised by pronounced membrane ruffling and actin cytoskeleton rearrangements that are accompanied by macropinocytosis and internalisation of the bacteria. Another cellular

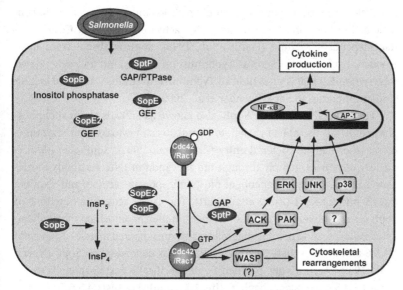

Figure 6.2. The action of *Salmonella* type III effectors SopE, SopE2, SopB, and SptP on RhoGTPases. SopE and SopE2 (closely related proteins) activate Cdc42 and Rac1 through their GEF activity. SopB also activates Cdc42 through its inositol phosphatase activity, but it is unclear how the SopB-mediated conversion of inositol pentakisphosphate (InsP$_5$) into inositol tetrakisphosphate (InsP$_4$) leads to Cdc42 activation. The interaction of the activated RhoGTPases with downstream effectors (ACK, PAK, and presumably WASP) mediates the *Salmonella*-induced cellular responses that promote bacterial entry and induce an inflammatory response. SptP, through its GAP activity towards Cdc42 and Rac1, seems to subsequently reverse *Salmonella*-induced cellular changes. To simplify the figure, SptP is represented promoting release of the GDP-bound Cdc42/Rac1 from membranes, which has never been experimentally demonstrated.

response is the activation of the mitogen-activated protein kinases (MAPKs), extracellular signal regulatory kinase (ERK), c-Jun NH$_2$-terminal kinase (JNK), and p38. Stimulation of these MAPK pathways leads to the activation of the transcription factors AP-1 and nuclear factor (NF)-κB, leading to the production of pro-inflammatory cytokines, an important feature of *Salmonella* pathogenesis. These cellular responses result mostly from the concerted activity of a subset of SPI-1 effectors (SopE, SopB and, SptP) acting on RhoGTPases (Figure 6.2).

The *Salmonella* SPI-1 type III effector, SopE, binds to Cdc42 and Rac1, and exhibits *in vitro* GEF activity for both GTPases (Figure 6.2) (Hardt et al., 1998). Consistently, transient transfection or microinjection of SopE into host cells results in the stimulation of Cdc42- and Rac1-dependent actin cytoskeleton rearrangements resembling those induced by *Salmonella* infection. Furthermore, SopE induces JNK activation, also in a Cdc42- and Rac1-dependent manner (Hardt et al., 1998). SopE is absent from most *S. enterica* subspecies I serovar Typhimurium strains, but a closely related protein (SopE2) that is present in all Typhimurium strains was found to have similar properties to SopE (Stender et al., 2000).

The SPI-1 TTSS effector SopB (also known as SigD) is an inositol phosphatase that is also able, by itself, to stimulate actin cytoskeleton rearrangements and mediate bacterial entry (Zhou et al., 2001). SopB also induces nuclear responses, mainly through the activation of JNK. Its ability to mediate bacterial entry is dependent on its phosphatase activity and requires Cdc42, but not Rac1 (Zhou et al., 2001). The Cdc42-activating functions of SopB are most likely the result of changes in phosphoinositide metabolism. *Salmonella* infection of intestinal cells results in a marked increase in inositol 1,4,5,6-tetrakisphosphate [Ins(1,4,5,6)P$_4$] that is dependent on SopB (Norris et al., 1998). Accordingly, purified SopB specifically desphosphorylates inositol 1,3,4,5,6-pentakisphosphate [Ins(1,3,4,5,6)P$_5$] to Ins(1,4,5,6)P$_4$ *in vitro* (Zhou et al., 2001). How the SopB-mediated conversion of Ins(1,3,4,5,6)P$_5$ to Ins(1,4,5,6)P$_4$ activates Cdc42 is unknown (Figure 6.2). The process of *Salmonella* entry is also modulated by two other SPI-1 TTSS substrates, SipA and SipC, which directly modulate actin dynamics through binding to actin (Galán, 2001).

The actin cytoskeleton changes induced by *Salmonella* are reversible, and after bacterial invasion the infected cells regain their normal architecture (Galán, 2001). The SPI-1 type III effector SptP seems to actively participate in this process. The SptP protein has a two-domain modular architecture. Accordingly, SptP possesses two distinct biochemical activities. The amino-terminal shows GAP activity towards Cdc42 and Rac1 (Fu and Galán, 1999), and the carboxyl-terminal domain exhibits tyrosine phosphatase activity (Kaniga et al., 1996). The rebuilding of the normal architecture of the host cell actin cytoskeleton that follows *Salmonella* entry appears to be mediated entirely by the GAP domain of SptP (Fu and Galán, 1999), which presumably reverses the activation of RhoGTPases by SopE (Figure 6.2).

Thus, the concerted action of SopE and SptP promotes bacterial internalisation through the GEF activity of SopE, which is followed by the reestablishment of the normal cytoskeleton architecture via the GAP activity of SptP. The cellular basis for the implicit temporal regulation in SopE and SptP activity has recently been shown to be due to differential host cell proteasome-mediated degradation kinetics of these two type III effectors (Kubori and Galán, 2003).

Modulation of RhoGTPase Signalling by *Shigella* Effectors

Upon contact with cultured epithelial cells, *Shigella* also induces the formation of membrane leaflets that rise and merge above the bacterial body to
allow its internalisation by the cell in a macropinocytic process. This process
is determined by actin polymerisation at the site of bacterial contact with
the cell membrane and involves the RhoGTPases, Cdc42, Rac1, and RhoA
(reviewed by Tran Van Nhieu et al., 2000).

The IpaB and IpaC proteins form a pore complex in the host cell membrane, presumably through which the other *Shigella* type III effectors are
delivered inside the eukaryotic cell (Tran Van Nhieu et al., 2000). The
IpaB – IpaC complex is also required for *Shigella* entry, as first suggested
by the observation that epithelial cells internalise latex beads coated with
IpaB – IpaC (Menard et al., 1996). The role of IpaB – IpaC in internalisation was further strengthened by addition of IpaC to semi-permeabilised
cells or microinjection of IpaC in intact cells, which in both cases induces
the formation of filopodial and lamellipodial extensions, resembling those
induced by *Shigella* (Van Nhieu et al., 1999). The effects of IpaC on the
cytoskeleton are most likely to be mediated by activation of Cdc42 (Van
Nhieu et al., 1999). Because Cdc42 can activate Rac1 (Hall, 1998), this allows the conversion of the IpaC-induced filopodial structures into lamellipodia. The IpaC protein is attached to the plasma membrane, presumably through a central hydrophobic domain, and the amino- and carboxy-
termini of IpaC are believed to mediate the described actin rearrangements
(Tran Van Nhieu et al., 2000). However, IpaC does not possess GEF activity, and the mechanism by which it activates Cdc42 and subsequently
Rac1 is unknown. Another *Shigella* effector, VirA, has been shown to be
required for efficient entry of *Shigella* into epithelial cells (Uchiya et al.,
1995). VirA interacts with tubulin to promote microtubule destabilisation
and elicit protrusions of membrane ruffling through the activation of Rac1,
thus promoting bacterial entry (Yoshida et al., 2002). After the activation of
the RhoGTPases, their downregulation is also required for completion of
bacterial internalisation. The IpaA protein, another type III effector, seems
to be involved in this process through binding to vinculin (Bourdet-Sicard
et al., 1999). Furthermore, as mentioned before, RhoA is also involved in
Shigella uptake. RhoA allows the recruitment at entry foci of cytoskeletal
proteins, such as ezrin, that is important for the organisation of the
IpaC-induced extensions into a productive entry site (Skoudy et al., 1999).
However, to date, no type III effector from *Shigella* has been shown to
activate RhoA.

Type III Effectors Acting on Small GTP-Binding Proteins and Preventing Bacterial Uptake

Pathogenic *Yersinia* spp. (*Y. enterocolitica, Y. pestis* and *Y. pseudotuberculosis*) multiply extracellularly in lymphatic tissues of their host. The *Yersinia* survival mechanism is to avoid the innate immune system, in particular by inhibiting phagocytosis and downregulating the anti-inflammatory response (reviewed by Cornelis, 2002). *Pseudomonas aeruginosa* is an opportunistic pathogen that is associated with acute infections when normal host defences are impaired or when extensive tissue damage has occurred (Lyczak et al., 2000). *P. aeruginosa* is also an extracellular pathogen and, although not as well established as *Yersinia*, a central feature of its pathogenicity seems to be to avoid being phagocytosed. Both *Yersinia* and *P. aeruginosa* inject into host cells type III effectors that disrupt RhoGTPases signalling pathways that are known to play an important role in phagocytosis processes (Caron and Hall, 1998). In addition, one of the type III effectors delivered by *P. aeruginosa* also acts on small GTP-binding proteins of the Ras and Rab family, which may confer upon *P. aeruginosa* the capacity to inhibit wound healing processes, tissue regeneration, and motility, thus compromising cell viability (Olson et al., 1999).

The Activity of *Yersinia* YopE and YopT on RhoGTPases

Four *Yersinia* type III toxins, YopE, YopH, YpkA (YopO in *Y. enterocolitica*), and YopT, have been shown to confer resistance to phagocytosis by macrophages and polymorphonuclear leukocytes (PMNs) (Figure 6.3) (Grosdent et al., 2002). Although it interacts with RhoA and Rac1, the action of YpkA/YopO on RhoGTPases has not been clearly established. For this reason, YpkA/YopO will be discussed separately below.

Delivery of YopE into epithelial cells and macrophages leads to a cytotoxic response characterised by cell rounding and detachment from the extracellular matrix, resulting from the disruption of the actin microfilament network (Rosqvist et al., 1991). YopE is similar to the amino terminal domain of SptP. It displays *in vitro* GAP activity towards RhoA, Rac1, and Cdc42 and bears an arginine-finger motif similar to those found in mammalian GAP proteins (von Pawel-Rammingen et al., 2000). The mechanism of inhibition of phagocytosis by YopE results from its GAP activity (Black and Bliska, 2000). The actual preferred substrate(s) of YopE under physiological conditions seems to be Rac1 (Figure 6.3) (Andor et al., 2001).

YopT exerts a strong depolymerising effect on actin (Iriarte and Cornelis, 1998). This effector is a cysteine protease that releases RhoA from the cell

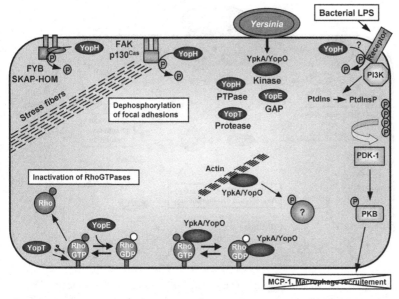

Figure 6.3. The action of the *Yersinia* type III effectors YopE, YopH, YopT, and YpkA/YopO. Through their respective GAP and protease activities, YopE and YopT inactivate RhoGTPases. The YpkA/YopO kinase interacts with, and is activated by, actin. YpkA/YopO also interacts with RhoGTPases (either GTP or GDP bound), but the target(s) of the kinase is unknown. The PTPase YopH is targeted to focal adhesions and to other protein complexes, where it dephosphorylates proteins such as focal adhesion kinase (FAK), p130^{Cas}, Fyb, and SKAP-HOM. YopH also blocks the PI3K/PKB pathway, probably by acting on a tyrosine-phosphorylated receptor. Upon stimulation with *Yersinia* LPS, the activated PI3K phosphorylates inositol phospholipids (PtdIns). The phosphorylated PtdIns (PtdInsP) then recruit proteins, such as phosphoinositide-dependent-kinase-1 (PDK-1), that phosphorylate and activate PKB. The production of MCP-1 is dependent on this phosphorylation cascade, and thus YopH should prevent macrophage recruitment.

membrane by cleavage of isoprenylated RhoA near its carboxyl termini (Shao et al., 2002; Shao et al., 2003; Zumbihl et al., 1999). *In vitro*, YopT also releases Rac and Cdc42 from membranes by an identical mechanism (Figure 6.3) (Shao et al., 2002; Shao et al., 2003), but bacterially translocated YopT seems to act only on RhoA (Aepfelbacher et al., 2003)

P. aeruginosa ExoS and ExoT, Inhibition of Phagocytosis, and Wound Healing Processes

At least four toxins, Exoenzyme S (ExoS), ExoT, ExoY, and ExoU, are delivered into the eukaryotic host cell cytoplasm by the TTSS of *P. aeruginosa*. Two of

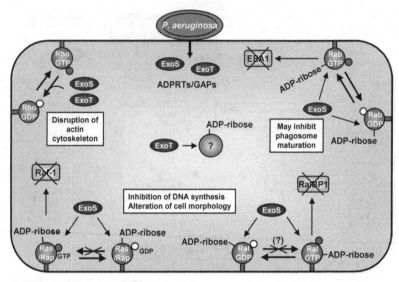

Figure 6.4. The action of *P. aeruginosa* ExoS and ExoT on small GTP-binding proteins. ExoS and ExoT are highly similar proteins exhibiting both ADPRT and GAP activity. ExoS ADPRT activity targets the small GTP-binding proteins H-Ras (Ras in the figure), Rap1B (Rap), RalA (Ral), and Rab5 (Rab). In the case of Ras and Rap (Ras/Rap in the figure), ADP-ribosylation by ExoS inhibits the interaction of the Ras proteins with their GEFs, and this prevents the interaction with downstream effectors such as Raf-1. In the case of Ral and Rab, ADP-ribosylation by ExoS has been shown to inhibit their interaction with downstream effectors RalBP1 and EEA1. ADP-ribosylation of Ras may also prevent the activity of RalGDS, a specific GEF for Ral, and, thus, interfere with GEF-catalyzed RalGDP to RalGTP exchange (Fraylick et al., 2002). The cellular target of the ADPRT activity of ExoT (0.2–1% of that of ExoS) is unknown. The GAP activity of both ExoS and ExoT is displayed towards RhoGTPases, such as Cdc42, Rac1, and RhoA.

them, ExoS and ExoT, exert their effects on signalling mediated by small GTP-binding proteins (Figure 6.4).

ExoS is a bifunctional protein containing an amino-terminal domain that possesses GAP activity (Goehring et al., 1999), and a carboxyl-terminal domain displaying ADP-ribosyltransferase (ADPRT) activity (Frank, 1997). The action of ExoS on epithelial cells causes a decrease in DNA synthesis, long-term alterations in cell morphology, microvillus effacement, and loss of the ability to re-adhere to plastic surfaces (Olson et al., 1999), thus affecting cell viability. These effects seem to be largely due to the ADPRT activity of ExoS (Fraylick et al., 2001). *In vitro*, this ADPRT activity is dependent on a eukaryotic protein termed FAS (factor activating ExoS) and is exerted preferentially on small GTP-binding proteins (Frank, 1997). *In vivo*, ExoS has been shown to ADP-ribosylate Ras (H-Ras, Rap1B, and RalA) and Rab proteins (Figure 6.4) (Barbieri et al., 2001; Fraylick et al., 2002; McGuffie et al., 1998; Riese et al., 2001). ExoS ADP-ribosylation of H-Ras inhibits the interaction with its specific GEF, Cdc25, and consequently inhibits the nucleotide exchange reaction catalysed by Cdc25 (Ganesan et al., 1999). Because only GTP-bound Ras is capable of interacting with its downstream effectors, this disrupts Ras-mediated signal transduction (Ganesan et al., 1998). Accordingly, *in vivo* ExoS-mediated ADP-ribosylation of H-Ras inhibits the ability of the GTPase to interact with its downstream effector Raf-1 (Vincent et al., 1999). The modification of H-Ras by ExoS correlates directly with its ability to inhibit DNA synthesis (McGuffie et al., 1998).

ExoS ADP-ribosylation of Rap1B also inhibits the interaction between Rap1B and its GEF, C3G, which could modulate the oxidative-burst in neutrophils or signal transduction through integrin-mediated pathways (Riese et al., 2001). ExoS ADP-ribosylation of RalA alters its ability to bind its downstream effector RalBP1 (Fraylick et al., 2002). Similarly, the action of ExoS on Rab5 inhibits the interaction of the GTPase with its effector EEA1 (early endosome antigen 1) (Barbieri et al., 2001), which is predicted to affect phagosome maturation. Therefore, ExoS may also inhibit phagocytosis through its action on Rab.

ExoS displays GAP activity towards RhoA, Rac1, and Cdc42, both *in vitro* and *in vivo* (Figure 6.4) (Goehring et al., 1999; Krall et al., 2002). Transfection of the amino terminal part of ExoS, or delivery of ExoS by the TTSS of *P. aeruginosa* in eukaryotic cells, disrupts actin filaments (Krall et al., 2002; Pederson et al., 1999). In addition, type III secretion delivered ExoS confers phagocytosis resistance against macrophages (Frithz-Lindsten et al., 1997).

The primary structure of ExoT is 75% identical to that of ExoS. However, the ADPRT activity of ExoT is only 0.2–1% of that of ExoS (Frank, 1997), and ExoT does not interfere with cell viability or with Ras signalling (Sundin et al., 2001). ExoT disrupts the cytoskeletal architecture of epithelial cells (Vallis et al., 1999), and inhibits the uptake of *P. aeruginosa* by epithelial cells and macrophages (Garrity-Ryan et al., 2000). These activities are partially dependent on the GAP activity that ExoT exhibits towards RhoA, Rac1, and Cdc42 both *in vitro* and *in vivo* (Figure 6.4) (Kazmierczak and Engel, 2002; Krall et al., 2000). Furthermore, the GAP activity of ExoT is required for the ability of the protein to inhibit lung epithelial wound repair *in vitro* (Geiser et al., 2001). Since point mutations at the arginine finger of ExoT result only in intermediate defects, it is predicted that its ADPRT activity may also contribute to cell rounding and anti-internalisation activities (Sundin et al.,

2001). As yet, the host protein(s) that ExoT ADP-ribosylates has not been identified.

The EPEC Effectors Tir and Map Also Modulate Small RhoGTPases

Central to enteropathogenic *E. coli* (EPEC) – mediated disease is its colonisation of the intestinal epithelium. EPEC reside extracellularly on the surface of the infected epithelial cell, where they deliver toxins by using a TTSS. At least four effector proteins are delivered by EPEC into target cells, Tir (translocated intimin receptor), Map (mitochondrial-associated protein), EspG, and EspF. A recent study indicates that Tir and Map modulate RhoGTPase signalling pathways.

Tir is by far the best studied of the EPEC effectors (reviewed by Goosney et al., 2000). After its delivery into the target cell, Tir is inserted into the plasma membrane, where it serves as a receptor for the bacterial outer membrane intimin. The interaction between Tir and intimin triggers signalling events that subvert the normal regulation of the host cell cytoskeleton and lead to the formation of the pedestal-like structures that characterise EPEC infection of epithelial cells. This process is not dependent on RhoGTPases, although Tir has a putative GAP motif (Kenny et al., 2002). However, EPEC binding to HeLa cells also induces a Tir-independent cytoskeletal rearrangement characterised by early and transient formation of filopodia-like structures at sites of infection (Kenny et al., 2002). Filopodia formation requires the Map effector, a protein that targets host mitochondria where it appears to disrupt membrane potential (Kenny and Jepson, 2000). Map-mediated filopodia formation is dependent on Cdc42, which is activated by an unknown mechanism, and is independent of Map targeting to the mitochondria (Kenny et al., 2002). Interestingly, in addition to the formation of the pedestal structures, Tir binding to intimin also downregulates filopodia formation (Kenny et al., 2002). This suggests that Tir, like SptP, YopE, ExoS, and ExoT, could indeed have a GAP activity. However, the consequences of the Tir and Map manipulation of Cdc42 activity for EPEC pathogenesis are unclear.

Type III – Delivered Kinases and Phosphatases Interfering with Signalling Pathways

Phosphorylation cascades play an important role in many animal signal transduction pathways. For example, the MAPK signalling pathways (Garrington and Johnson, 1999), the phosphatidylinositol 3-kinase (PI3K) pathway

(Vanhaesebroeck and Alessi, 2000), and the phosphorylation cascade that activates NF-κB (Karin and Lin, 2002) are each involved in the control of a multitude of cellular processes. These pathways are obvious targets for bacterial virulence factors. We have already described SopE from *Salmonella*, which acts on MAPK signalling pathways through its GEF activity on RhoGT-Pases (see above). However, in addition, some of the type III effectors are themselves kinases and phosphatases, thus interfering directly on host phosphorylation signalling cascades. Kinases and phosphatases delivered by TTSS into eukaryotic cells have been identified in *Salmonella*, *Shigella*, and *Yersinia*.

Yersinia YopH, a Tyrosine Phosphatase That Inhibits Innate and Adaptive Immune Responses

YopH is among the most powerful protein tyrosine phosphatases (PTPases) known and it has long been known to contribute to the ability of *Yersinia* to resist phagocytosis by macrophages (Rosqvist et al., 1988). When injected into J774 macrophages, YopH dephosphorylates p130^{Cas} and disrupts focal adhesions (Hamid et al., 1999). Other YopH targets in J774 macrophages are the Fyn-binding protein Fyb (Hamid et al., 1999) and the scaffolding protein SKAP-HOM (Figure 6.3) (Black et al., 2000). These two proteins interact with each other and become tyrosine phosphorylated in response to macrophage adhesion. It is likely that the action of YopH against focal adhesions, p130Cas, and Fyb, is relevant to the antiphagocytic action (Deleuil et al., 2003; Persson et al., 1999). YopH also suppresses the oxidative burst in macrophages (Bliska and Black, 1995), and protects against phagocytosis by PMNs (Grosdent et al., 2002; Visser et al., 1995).

Recent observations have shown that YopH also contributes to the downregulation of the inflammatory response (Figure 6.3) (Sauvonnet et al., 2002). Upon infection, macrophages release the monocyte chemoattractant protein 1 (MCP-1), a chemokine involved in the recruitment of other macrophages to the sites of infection. In fact, the MCP-1 mRNA levels are downregulated in macrophages infected with *Y. enterocolitica*, and this inhibition is dependent upon YopH. MCP-1 synthesis is known to be under the control of the PI3K pathway, which is involved in the control of multiple cellular processes (Vanhaesebroeck and Alessi, 2000). Consistently, with its negative action on MCP-1 mRNA levels, YopH was shown to abrogate the PI3K-dependent activation of protein kinase B (PKB, also called Akt) (Sauvonnet et al., 2002). The site of action of YopH on this cascade is still unknown but it is likely to be a tyrosine-phosphorylated receptor upstream from PI3K (Figure 6.3). Furthermore, YopH (together with YopE and YopJ/P) also counteracts YopB-stimulated pro-inflammatory signalling in infected epithelial cells, and is likely that this is dependent on its action on the PI3K pathway (Viboud et al., 2003). In addition, YopH also seems to incapacitate the host adaptive immune response. T and B cells transiently exposed to *Y. pseudotuberculosis* are impaired in their ability to be activated through their antigen receptors. YopH appears to be the main effector involved in this block of activation (Yao et al., 1999), which most likely also involves the PI3K/PKB pathway (Sauvonnet et al., 2002). In agreement with these observations, recombinant YopH introduced into human T lymphocytes was shown to dephosphorylate the primary signal transducer for the T cell receptor, the Lck tyrosine kinase (Alonso et al., 2004).

SptP, a PTPase Delivered by *Salmonella* SPI-1 TTSS

As already pointed out, the GAP domain of the SptP protein reverses the cytoskeleton changes induced by *Salmonella*. In addition, the bifunctional SptP protein appears to downregulate the *Salmonella*-induced activation of JNK and ERK (Fu and Galán, 1999; Murli et al., 2001). At least in the case of inhibiting ERK activation, this capacity seems to rely on the PTPase activity of SptP (Murli et al., 2001). Therefore, SptP may be a general downregulator of the cellular responses stimulated by *Salmonella*, and its PTPase activity may play an important role in this process. Clearly, the tyrosine phosphatase activity of SptP is not exerted on proteins phosphorylated by MAPKs, as these are serine/threonine protein kinases. To date, only the intermediate filament protein vimentin was identified as a potential target of the SptP PTPase activity (Murli et al., 2001). However, other target(s) of the PTPase remain to be identified, as it is unclear how the action of SptP on vimentin could contribute to reversing *Salmonella*-induced cellular responses.

SopB/SigD and IpgD, the Inositol Phosphatases from *Salmonella* and *Shigella*

As mentioned before, SopB displays inositol phosphatase activity both *in vitro* and *in vivo*. In addition to stimulating actin cytoskeleton reorganisation, SopB, through its inositol phosphatase activity, has been associated with the modulation of chloride secretion induced by *Salmonella* infection (Norris et al., 1998), and is known to be essential for the activation of PKB that is observed upon infection of epithelial cells by *Salmonella* (Steele-Mortimer et al., 2000). The modulation of chloride secretion, which directly contributes to diarrhoea, appears to be a consequence of products of inositol metabolism, in particular Ins(1,4,5,6)P$_4$ (Feng et al., 2001). With respect to PKB, the mechanism of activation by SopB and its importance in *Salmonella* pathogenesis are unclear. More recently, SopB was shown to be responsible for the disappearance of phosphatidylinositol (PtdIns)(4,5)P_2 at the invaginating regions of the ruffles induced by *Salmonella*. This reduces the rigidity of the membrane skeleton and is thought to promote the formation of the *Salmonella*-containing vacuoles (Terebiznik et al., 2002).

The *Shigella* effector IpgD is 59% similar and 41% identical to SopB at the amino acid level. IpgD has been shown to be involved in the modulation of the host cell response after contact of *Shigella* with epithelial cells (Niebuhr et al., 2000). Furthermore, purified IpgD displays PtdIns phosphatase activity *in vitro* and in infected cells (Niebuhr et al., 2000) (Marcus et al., 2001), and also activates PKB in an inositol phosphatase-dependent fashion (Marcus et al., 2001). *In vitro*, IpgD displays the greatest activity towards PtdIns(4,5)P_2, and the transformation of PtdIns(4,5)P_2 into PtdIns(5)P by IpgD is responsible for dramatic morphological changes of the host cell, which are thought to promote *Shigella* entry (Niebuhr et al., 2002).

YpkA/YopO, the *Yersinia* type III Delivered Protein Kinase

YpkA (for <u>Y</u>ersinia <u>p</u>rotein <u>k</u>inase <u>A</u>) (called YopO in *Y. enterocolitica*) is also an effector that modulates the dynamics of the cytoskeleton (Hakansson et al., 1996), and that contributes to *Yersinia* resistance to phagocytosis (Grosdent et al., 2002). It is an autophosphorylating serine-threonine kinase (Galyov et al., 1993), which shows some sequence and structural similarity to RhoA-binding kinases (Dukuzumuremyi et al., 2000), but which becomes active only after interacting with actin (Juris et al., 2000). In addition to being an activator of YpkA, actin can also function as an *in vitro* substrate of the kinase. YpkA/YopO interacts with, but does not phosphorylate, RhoA and Rac1 irrespective of the nucleotide bound, and apparently without affecting the guanine-nucleotide exchange capacity (Dukuzumuremyi et al., 2000). However, the kinase target and the exact mode of action of YpkA/YopO remain unknown (Figure 6.3).

The Modulation of Death Pathways and Inflammation by Type III Effectors

Several bacteria harbouring TTSS induce cell death in eukaryotic cells, particularly in macrophages. The process has been well studied in *Yersinia, Shigella*, and *Salmonella*, where the type III effectors YopJ/P, IpaB, and SipB, respectively, trigger cell death. Type III effectors from *P. aeruginosa* and EPEC were also reported to promote cell death, but in these cases the molecules and death pathways involved have not been thoroughly characterised. In the case of *Yersinia* and *Shigella*, it is clearly established that macrophage cell death occurs by apoptosis. However, while apoptosis is normally considered an immunologically silent death process unaccompanied by inflammation, *Shigella-* and *Salmonella-*induced macrophage cell death triggers characteristically acute inflammatory responses. In contrast, *Yersinia-*induced apoptosis is accompanied by a downregulation of the inflammatory response, at least in cell culture systems. Therefore, cell death induction by type III effectors, in addition to allowing the elimination of key immune cells and thus promoting bacterial proliferation, may be a way by which bacteria modulate the host inflammatory response.

The NF-κB transcription factor is central to the modulation of cell survival and immune and inflammatory responses (Karin and Lin, 2002). NF-κB exists in the cytoplasm of resting cells in an inactive form associated with inhibitory proteins termed inhibitor-kappa B (IκB). After cell stimulation, NF-κB activation is achieved through the phosphorylation of IκB by the IκB kinase (IKK) complex. The components of the IKK complex that regulate the NF-κB pathway include IKKα and IKKβ, which are activated by morphogenic and pro-inflammatory signals, respectively. Phosphorylated IκB is then selectively ubiquitinated, which targets its degradation by the proteasome. This allows NF-κB to translocate to the nucleus and activate transcription of genes involved in immune responses, including several cytokines.

Yersinia YopJ/P

YopJ (YopP in *Y. enterocolitica*) counteracts the normal pro-inflammatory response in various cell culture systems (Boland and Cornelis, 1998; Denecker et al., 2002; Schesser et al., 1998). This capacity results from the ability of YopJ/P to inhibit the MAPK pathways JNK, p38, and ERK1 and 2, and also the NF-κB signalling pathway (Figure 6.5). Furthermore, inhibition of the MAPK pathways abrogates phosphorylation of CREB, a transcription factor involved in the immune response (reviewed by Orth, 2002).

YopJ/P binds directly to the members of the superfamily of MAPK kinases (MKKs), blocking their phosphorylation and consequent activation (Orth et al., 1999). Similarly, YopJ/P interacts with IKKβ, presumably also preventing its phosphorylation and activation (Orth et al., 1999). Thus, YopJ/P acts by preventing activation of MKKs and IKK in the MAPK and NF-κB signalling pathways, respectively (Figure 6.5). The exact molecular mechanism by which YopJ/P acts is still elusive, but its secondary structure resembles that of an adenovirus cysteine protease. The presumed protease catalytic triad of YopJ/P is required for the ability of YopJ/P to inhibit the MAPK and NF-κB signalling pathways (Orth et al., 2000). Based on the observations that overexpression of YopJ/P results in a decrease of SUMOylated and ubiquitinated

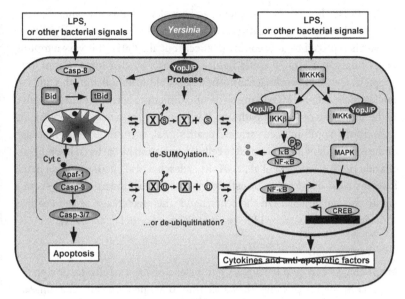

Figure 6.5. Action of *Yersinia* YopJ/P on NF-κB, MAPK, and apoptosis signalling pathways. YopJ/P binds to IKKβ and MKKs and block their activation through phosphorylation by upstream MKK kinases (MKKKs). This inhibits the activity of the NF-κB and CREB transcription factors, thus, preventing transcription of anti-apoptotic genes and of pro-inflammatory cytokines. YopJ/P also induces apoptosis in macrophages, either by directly activating a death pathway or indirectly by blocking the synthesis of anti-apoptotic factors together with bacterial LPS/ lipoproteins apoptotic signalling. The apoptotic cascade is most probably triggered by caspase-8 activation, leading to Bid cleavage. The subsequent translocation of tBid to the mitochondria induces the release of cytochrome *c*, which binds to the apoptotic protease activating factor-1 (Apaf-1) and leads to recruitment and activation of caspase-9 that, in turn, activates executioner caspases-3 and -7. The YopJ/P protease may act on SUMOylated and/or ubiquitinated proteins, but the relevance of these activities for YopJ/P function is unclear.

proteins, it was proposed that YopJ/P could be an ubiquitin or a SUMO protease (SUMO are ubiquitin-like modifiers that are involved in stabilisation or destabilisation of proteins) (Orth, 2002; Orth et al., 2000). It is unclear how a de-SUMOylating activity could contribute to the ability of YopJ/P to disrupt MAPK and NF-κB signalling, but the NF-κB pathway is regulated at two distinct points by ubiquitination (Orth, 2002).

YopJ/P induces apoptosis in macrophages, but not in other cell types (Mills et al., 1997; Monack et al., 1997). This apoptosis is accompanied by cleavage of the cytosolic protein Bid, the release of cytochrome *c* from the mitochondria, and the activation of caspase-3, -7, and -9 (Figure 6.5) (Denecker et al., 2001). The apoptosis-inducing activity of YopJ/P is lost when the cysteine residue forming the catalytic triad responsible for its putative protease activity is replaced by a threonine (Denecker et al., 2001). It is very tempting to speculate that YopJ/P could induce apoptosis by cleaving a pro- or anti-apoptotic factor. However, it has not been possible to demonstrate this and it is unclear how a de-SUMOylating or de-ubiquitinating activity could be related to the induction of a death pathway. It is even unclear whether apoptosis results from a YopP-induced early cell death signal, or from the YopP-blockage of the NF-κB signalling, coupled to cellular activation by bacterial lipopolysaccharide (LPS) (Denecker et al., 2001; Ruckdeschel et al., 2001). On one hand, it seems clear that YopJ/P transfected macrophages undergo apoptosis in a significant percentage. On the other hand, this percentage of cell death is considerably increased by LPS treatment (Ruckdeschel et al., 2001).

Shigella IpaB

Macrophages can engulf *Shigella*, but instead of successfully destroying the bacteria in the phagosome, they succumb to apoptotic death. The *Shigella*-induced macrophage apoptosis has been described both *in vitro* and *in vivo* (Zychlinsky et al., 1992; Zychlinsky et al., 1996). The IpaB type III-delivered toxin is necessary and sufficient to induce apoptosis (Chen et al., 1996b; Zychlinsky et al., 1994). Following the escape of *Shigella* from the macrophage phagosome, IpaB is secreted into the cytoplasm where it binds to caspase-1. Furthermore, *in vivo* studies using different knockout mice showed that IpaB-induced apoptosis is absolutely dependent on caspase-1 activation (Hilbi et al., 1998). IpaB-induced caspase-1–dependent apoptosis of macrophages triggers an acute inflammatory response, characterised by a massive influx of PMNs that infiltrate the infected site and destabilise the epithelium (Sansonetti, 2001). Caspase-1 is also known as interleukin (IL)-1-converting enzyme, due to its ability to cleave pro-IL-1β and the related cytokine pro-IL-18 into their mature active forms. Accordingly, IpaB-activated caspase-1 can cleave the two cytokines, which are then released from the dying macrophage (Hilbi et al., 1998). Therefore, *Shigella*-induced macrophage cell death is central to the early triggering of inflammation. This inflammatory response causes significant tissue destruction and facilitates further invasion by the bacteria, but inflammation is ultimately responsible for controlling the infection (Sansonetti, 2001).

Salmonella SipB

In vitro studies indicate that *Salmonella* spp. induces cell death in macrophages (Chen et al., 1996a; Lindgren et al., 1996; Monack et al., 1996). The process by which *Salmonella* induces cell death is not totally clear, as features of both apoptotic and necrotic death have been described (Boise and Collins, 2001). The TTSS SPI-1 SipB protein is highly similar to the *Shigella* IpaB protein, and SipB seems to promote rapid cell death by a mechanism resembling that of IpaB (Hersh et al., 1999). However, macrophages from *casp-1* knockout mice are still susceptible to *Salmonella*-induced cell death in a SipB-dependent way, although with delayed kinetics (Jesenberger et al., 2000). Furthermore, *Salmonella* also kill macrophages independently of SipB and of SPI-1 TTSS, most likely by mechanisms involving the SPI-2 TTSS (van der Velden et al., 2000), but the bacterial toxin(s) involved remain to be identified. Quite recently, it was suggested that SipB promotes autophagy and cell death by disrupting mitochondria (Hernandez et al., 2003).

P. aeruginosa ExoU, ExoS, PcrV, PopB, and PopD

The delivery of ExoU by *P. aeruginosa* TTSS into host cells induces a cytotoxic phenotype in tissue culture models and a fatal outcome in an acute lung infection model (Finck-Barbancon et al., 1997). The cell death mechanism mediated by ExoU in macrophages and epithelial cells has the characteristics of necrosis rather than apoptosis (Hauser and Engel, 1999). Recently, it was shown that ExoU is a phospholipase and that its cytotoxic action can be blocked by different phospholipase A2 inhibitors (Phillips et al., 2003; Sato et al., 2003). In addition, *P. aeruginosa* induces apoptosis in macrophages and epithelial cells in an ExoU-independent, but TTSS-dependent fashion, and *P. aeruginosa* clinical isolates that do not express ExoS or ExoY also induce apoptosis (Hauser and Engel, 1999). In contrast to this observation, *P. aeruginosa*-mediated apoptosis of epithelial and fibroblast cell lines was shown to require the ADPRT activity of ExoS (Kaufman et al., 2000). In addition, the pore forming activity of the translocators PcrV, PopB, and PopD has been reported to promote the death of macrophages and neutrophils (Dacheux et al., 2001).

Enteropathogenic *E. coli* EspF

EPEC induces cell death in epithelial host cells by a process that resembles apoptosis in some ways, but not in others (Crane et al., 1999). The EPEC type III effector EspF disrupts host intestinal barrier function, thus contributing to diarrhoea (McNamara et al., 2001), and in addition it appears to mediate cell death (Crane et al., 2001). Because cell death promoted by EspF has features of pure apoptosis (Crane et al., 2001), this indicates that EPEC may also trigger cell death independently of EspF. To date, the apoptotic pathway induced by EspF is unknown.

Type III Effectors Affecting Other Host Signalling Pathways or with Unknown Function

In addition to the type III effectors described above, others have been identified. These include the *Salmonella* SPI-2 TTSS effectors SifA, SseG, SseJ, and SpiC, which manipulate host signalling to promote intracellular growth (reviewed by Waterman and Holden, 2003; Salcedo and Holden, 2003); adenylate cyclase ExoY from *P. aeruginosa* (Ruiz-Albert et al., 2002); a number of leucine-rich repeat (LRR) proteins that seem to be translocated into the nucleus of infected cells and are essential for the pathogenesis of *Yersinia* (YopM), *Shigella* (IpaHs), or *Salmonella* (SspHs and SlrP) (Cornelis, 2002; Galán, 2001; Sansonetti, 2001). The function of these LRR type III effector proteins is mostly mysterious, but new insights came from two recent reports. *Yersinia* YopM has been shown to interact directly with, and stimulate the activity of, protein kinase C-like 2 (PRK2) and ribosomal S6 protein kinase 1 (RSK1) (McDonald et al., 2003); *Salmonella* SspH1 and *Shigella* IpaH9.8 were reported to inhibit NF-κB-dependent gene expression (Haraga and Miller, 2003). Other type III effector proteins await further characterisation, for example, *Salmonella* AvrA, SopA, and SopD (Galán, 2001), and EPEC EspG (Elliot et al., 2001).

CONCLUSIONS

Remarkable progress has been made in the last few years in understanding the mode of action of proteins delivered into eukaryotic cells by TTSSs. These studies have outlined the diversity of functions and biochemical activities of type III effectors and have revealed a new class of bacterial toxins distinguished by the way they enter the host cell. In several cases these proteins seem to mimic host proteins, or to have counterparts in the eukaryotic world, thus suggesting that they were acquired through horizontal transfer during the co-evolution of bacterial pathogens and their host organisms. In this respect, we await the characterisation of type III effectors from bacterial pathogens that reside exclusively inside host cells, such as *Chlamydia* spp., where a more intimate cross-talk has developed. In addition, the ongoing efforts to better understand the action of the currently known type III effectors promises, as always, not only to help in the understanding of bacterial pathogenesis, but also to provide us with priceless tools for studying the functioning of the eukaryotic cell.

ACKNOWLEDGEMENTS

The laboratory of G. R. C. is supported by the Swiss National Science Founda-
tion (contract Nr 32–65393.01). The laboratory is also member of a European
Union network (HPRN-CT-2000–00075). L. J. M. is supported by a post-
doctoral fellowship (SFRH/BPD/3582/2000) from Fundação para a Ciência
e Tecnologia (Portugal).

REFERENCES

Aepfelbacher M, Trasak C, Wilharm G, Wiedemann A, Trulzsch K, Krauss K,
Gierschik P, and Heesemann J (2003). Characterization of YopT effects
on Rho GTPases in *Yersinia enterocolitica*-infected cells. *J. Biol. Chem.*, **278**,
33217–33223.

Alonso A, Bottini N, Bruckner S, Rahmouni S, Williams S, Schoenberger S P,
and Mustelin T (2004). Lck dephosphorylation at Tyr394 and inhibition of
T cell antigen receptor signalling by *Yersinia* phosphatase YopH. *J. Biol.
Chem.* **279**, 4922–4928.

Ador A, Trulzsch K, Essler M, Roggenkamp A, Wiedemann A, Heesemann J, and
Aepfelbacher M (2001). YopE of *Yersinia*, a GAP for Rho GTPases, selectively
modulates Rac-dependent actin structures in endothelial cells. *Cell Microbiol.*,
3, 301–310.

Barbieri A M, Sha Q, Bette-Bobillo P, Stahl P D, and Vidal M (2001). ADP-
ribosylation of Rab5 by ExoS of *Pseudomonas aeruginosa* affects endocytosis.
Infect. Immun., **69**, 5329–5234.

Black D S and Bliska J B (2000). The RhoGAP activity of the *Yersinia pseudotuberculosis* cytotoxin YopE is required for antiphagocytic function and virulence.
Mol. Microbiol., **37**, 515–527.

Black D S, Marie-Cardine A, Schraven B, and Bliska J B (2000). The *Yersinia*
tyrosine phosphatase YopH targets a novel adhesion-regulated signalling
complex in macrophages. *Cell. Microbiol.*, **2**, 401–414.

Bliska J B and Black D S (1995). Inhibition of the Fc receptor-mediated oxidative
burst in macrophages by the *Yersinia pseudotuberculosis* tyrosine phosphatase.
Infect Immun., **63**, 681–685.

Boise L H and Collins C M (2001). *Salmonella*-induced cell death: Apoptosis,
necrosis or programmed cell death? *Trends. Microbiol.*, **9**, 64–67.

Boland A and Cornelis G R (1998). Role of YopP in suppression of tumor necrosis
factor alpha release by macrophages during *Yersinia* infection. *Infect Immun.*,
66, 1878–1884.

Boquet P (2000). Small GTP binding proteins and bacterial virulence. *Microbes Infect.*, **2**, 837–843.

Bourdet-Sicard R, Rudiger M, Jockusch B M, Gounon P, Sansonetti P J, and Nhieu G T (1999). Binding of the *Shigella* protein IpaA to vinculin induces F-actin depolymerization. *EMBO J.*, **18**, 5853–5862.

Buttner D and Bonas U (2002). Port of entry – the type III secretion translocon. *Trends Microbiol.*, **10**, 186–192.

Caron E and Hall A (1998). Identification of two distinct mechanisms of phago-cytosis controlled by different Rho GTPases. *Science*, **282**, 1717–1721.

Chen L M, Kaniga K, and Galán J E (1996a). *Salmonella* spp. are cytotoxic for cultured macrophages. *Mol Microbiol.*, **21**, 1101–1115.

Chen Y, Smith M R, Thirumalai K, and Zychlinsky A (1996b). A bacterial invasin induces macrophage apoptosis by binding directly to ICE. *EMBO J.*, **15**, 3853–60.

Cornelis G R (2002). *Yersinia* type III secretion: send in the effectors. *J. Cell Biol.*, **158**, 401–408.

Cornelis G R and Van Gijsegem F (2000). Assembly and function of type III secretory systems. *Annu. Rev. Microbiol.*, **54**, 735–774.

Crane J K, Majumdar S, and Pickhardt D P (1999). Host cell death due to en-teropathogenic *Escherichia coli* has features of apoptosis. *Infect. Immun.*, **67**, 2575–2584.

Crane J K, McNamara B P, and Donnenberg M S (2001). Role of EspF in host cell death induced by enteropathogenic *Escherichia coli*. *Cell. Microbiol.*, **3**, 197–211.

Dacheux D, Goure J, Chabert J, Usson Y, and Attree I (2001). Pore-forming activity of type III system-secreted proteins leads to oncosis of *Pseudomonas aeruginosa*-infected macrophages. *Mol. Microbiol.*, **40**, 76–85.

Deleuil F, Mogemark L, Francis M S, Wolf-Watz H, and Fallman M (2003). In-teraction between the *Yersinia* protein tyrosine phosphatase YopH and eu-karyotic Cas/Fyb is an important virulence mechanism. *Cell. Microbiol.*, **5**, 53–64.

Denecker G, Declercq W, Geuijen C A, Boland A, Benabdillah R, van Gurp M, Sory M P, Vandenabeele P, and Cornelis G R (2001). *Yersinia enterocolitica* YopP-induced apoptosis of macrophages involves the apoptotic signalling cascade upstream of Bid. *J. Biol. Chem.*, **276**, 19706–19714.

Denecker G, Totemeyer S, Mota L J, Troisfontaines P, Lambermont I, Youta C, Stainier I, Ackermann M, and Cornelis G R (2002). Effect of low- and high-virulence *Yersinia enterocolitica* strains on the inflammatory response of human umbilical vein endothelial cells. *Infect. Immun.*, **70**, 3510–3520.

Dukuzumuremyi J M, Rosqvist R, Hallberg B, Akerstrom B, Wolf-Watz H, and Schesser K (2000). The *Yersinia* protein kinase A is a host factor inducible RhoA/Rac-binding virulence factor. *J. Biol. Chem.*, **275**, 35281–35290.

Elliot S J, Krejany E O, Mellies J L, Robins-Browne R M, Sasakawa C, and Kaper J B (2001). EspG, a novel type III system-secreted protein from enteropathogenic *Escherichia coli* with similarities to VirA of *Shigella flexneri*. *Infect. Immun.*, **69**, 4027–4033.

Feng Y, Wente S R, and Majerus P W (2001). Overexpression of the inositol phosphatase SopB in human 293 cells stimulates cellular chloride influx and inhibits nuclear mRNA export. *Proc. Natl. Acad. Sci. USA*, **98**, 875–879.

Finck-Barbancon V, Goranson J, Zhu L, Sawa T, Wiener-Kronish J P, Fleiszig S M, Wu C, Mende-Mueller L, and Frank D W (1997). ExoU expression by *Pseudomonas aeruginosa* correlates with acute cytotoxicity and epithelial injury. *Mol. Microbiol.*, **25**, 547–557.

Frank D W (1997). The exoenzyme S regulon of *Pseudomonas aeruginosa*. *Mol. Microbiol.*, **26**, 621–629.

Fraylick J E, LaRocque J R, Vincent T S, and Olson J C (2001). Independent and coordinate effects of ADP-ribosyltransferase and GTPase-activating activities of exoenzyme S on HT-29 epithelial cell function. *Infect. Immun.*, **69**, 5318–5328.

Fraylick J E, Riese M J, Vincent T S, Barbieri J T, and Olson J C (2002). ADP-ribosylation and functional effects of *Pseudomonas* Exoenzyme S on cellular RalA. *Biochem.*, **41**, 9680–9687.

Frithz-Lindsten E, Du Y, Rosqvist R, and Forsberg A (1997). Intracellular targeting of exoenzyme S of *Pseudomonas aeruginosa* via type III-dependent translocation induces phagocytosis resistance, cytotoxicity and disruption of actin microfilaments. *Mol. Microbiol.*, **25**, 1125–1139.

Fu Y and Galán J E (1999). A *Salmonella* protein antagonizes Rac-1 and Cdc42 to mediate host-cell recovery after bacterial invasion. *Nature*, **401**, 293–297.

Galán J E (2001). *Salmonella* interactions with host cells: Type III secretion at work. *Annu. Rev. Cell Dev. Biol.*, **17**, 53–86.

Galyov E E, Hakansson S, Forsberg A, and Wolf-Watz H (1993). A secreted protein kinase of *Yersinia pseudotuberculosis* is an indispensable virulence determinant. *Nature*, **361**, 730–732.

Ganesan A K, Frank D W, Misra R P, Schmidt G, and Barbieri J T (1998). *Pseudomonas aeruginosa* exoenzyme S ADP-ribosylates Ras at multiple sites. *J. Biol. Chem.*, **273**, 7332–7337.

Ganesan A K, Vincent T S, Olson J C, and Barbieri J T (1999). *Pseudomonas aeruginosa* exoenzyme S disrupts Ras-mediated signal transduction by inhibiting guanine nucleotide exchange factor-catalyzed nucleotide exchange. *J. Biol. Chem.*, **274**, 21823–21829.

Garrington T P and Johnson G L (1999). Organization and regulation of mitogen-activated protein kinase signalling pathways. *Curr. Opin. Cell Biol.*, **11**, 211–218.

Garrity-Ryan L, Kazmierczak B, Kowal R, Comolli J, Hauser A, and Engel J N (2000). The arginine finger domain of ExoT contributes to actin cytoskeleton disruption and inhibition of internalization of *Pseudomonas aeruginosa* by epithelial cells and macrophages. *Infect. Immun.*, **68**, 7100–7113.

Geiser T K, Kazmierczak B I, Garrity-Ryan L K, Matthay M A, and Engel J N (2001). *Pseudomonas aeruginosa* ExoT inhibits in vitro lung epithelial wound repair. *Cell. Microbiol.*, **3**, 223–236.

Goehring U M, Schmidt G, Pederson K J, Aktories K, and Barbieri J T (1999). The N-terminal domain of *Pseudomonas aeruginosa* exoenzyme S is a GTPase-activating protein for Rho GTPases. *J. Biol. Chem.*, **274**, 36369–36372.

Goosney D L, Gruenheid S, and Finlay B B (2000). Gut feelings: Enteropathogenic *E. coli* (EPEC) interactions with the host. *Annu. Rev. Cell Dev. Biol.*, **16**, 173–189.

Grosdent N, Maridonneau-Parini I, Sory M P, and Cornelis G R (2002). Role of Yops and adhesins in resistance of *Yersinia enterocolitica* to phagocytosis. *Infect. Immun.*, **70**, 4165–4176.

Hakansson S, Galyov E E, Rosqvist R, and Wolf-Watz H (1996). The *Yersinia* YpkA Ser/Thr kinase is translocated and subsequently targeted to the inner surface of the HeLa cell plasma membrane. *Mol. Microbiol.*, **20**, 593–603.

Hall A (1998). Rho GTPases and the actin cytoskeleton. *Science*, **279**, 509–514.

Hamid N, Gustavsson A, Andersson K, McGee K, Persson C, Rudd C E, and Fallman M (1999). YopH dephosphorylates Cas and Fyn-binding protein in macrophages. *Microb. Pathogenesis*, **27**, 231–242.

Haraga A and Miller S I (2003). A *Salmonella enterica* serovar Typhimurium translocated leucine-rich repeat effector protein inhibits NF-κB-dependent gene expression. *Infect. Immun.*, **71**, 4052–4058.

Hardt W D, Chen L M, Schuebel K E, Bustelo X R, and Galán J E (1998). *S. typhimurium* encodes an activator of Rho GTPases that induces membrane ruffling and nuclear responses in host cells. *Cell*, **93**, 815–826.

Hauser A R and Engel J N (1999). *Pseudomonas aeruginosa* induces type-III-secretion-mediated apoptosis of macrophages and epithelial cells. *Infect. Immun.*, **67**, 5530–5537.

Hernandez L D, Pypaert M, Flavell R A, and Galán J E (2003). A *Salmonella* protein causes macrophage cell death by inducing autophagy. *J. Cell. Biol.*, **163**, 1123–1131.

Hersh D, Monack D M, Smith M R, Ghori N, Falkow S, and Zychlinsky A (1999). The *Salmonella* invasin SipB induces macrophage apoptosis by binding to caspase-1. *Proc. Natl. Acad. Sci. USA*, **96**, 2396–2401.

Hilbi H, Moss J E, Hersh D, Chen Y, Arondel J, Banerjee S, Flavell R A, Yuan J, Sansonetti P J, and Zychlinsky A (1998). *Shigella*-induced apoptosis is dependent on caspase-1 which binds to IpaB. *J. Biol. Chem.*, **273**, 32895–32900.

Iriarte M and Cornelis G R (1998). YopT, a new *Yersinia* Yop effector protein, affects the cytoskeleton of host cells. *Mol. Microbiol.*, **29**, 915–929.

Jesenberger V, Procyk K J, Yuan J, Reipert S, and Baccarini M (2000). *Salmonella*-induced caspase-2 activation in macrophages: A novel mechanism in pathogen-mediated apoptosis. *J. Exp. Med.*, **192**, 1035–1046.

Journet L, Agrain C, Brosz P, and Cornelis G R (2003). The needle length of bacterial injectisomes is determined by a molecular ruler. *Science*, **302**, 1757–1760.

Juris S J, Rudolph A E, Huddler D, Orth K, and Dixon J E (2000). A distinctive role for the *Yersinia* protein kinase: Actin binding, kinase activation, and cytoskeleton disruption. *Proc. Natl. Acad. Sci. USA*, **97**, 9431–9436.

Kaniga K, Uralil J, Bliska J B, and Galán J E (1996). A secreted protein tyrosine phosphatase with modular effector domains in the bacterial pathogen *Salmonella* typhimurium. *Mol. Microbiol.*, **21**, 633–641.

Karin M and Lin A (2002). NF-κB at the crossroads of life and death. *Nat. Immunol.*, **3**, 221–227.

Kaufman M R, Jia J, Zeng L, Ha U, Chow M, and Jin S (2000). *Pseudomonas aeruginosa* mediated apoptosis requires the ADP-ribosylating activity of ExoS. *Microbiology*, **146**, 2531–2541.

Kazmierczak B I and Engel J N (2002). *Pseudomonas aeruginosa* ExoT acts in vivo as a GTPase-activating protein for RhoA, Rac1, and Cdc42. *Infect. Immun.*, **70**, 2198–2205.

Kenny B, Ellis S, Leard A D, Warawa J, Mellor H, and Jepson M A (2002). Co-ordinate regulation of distinct host cell signalling pathways by multifunctional enteropathogenic *Escherichia coli* effector molecules. *Mol. Microbiol.*, **44**, 1095–1107.

Kenny B and Jepson M (2000). Targeting of an enteropathogenic *Escherichia coli* (EPEC) effector protein to host mitochondria. *Cell. Microbiol.*, **2**, 579–90.

Krall R, Schmidt G, Aktories K, and Barbieri J T (2000). *Pseudomonas aeruginosa* ExoT is a Rho GTPase-activating protein. *Infect. Immun.*, **68**, 6066–6068.

Krall R, Sun J, Pederson K J, and Barbieri J T (2002). In vivo Rho GTPase-activating protein activity of *Pseudomonas aeruginosa* cytotoxin ExoS. *Infect. Immun.*, **70**, 360–367.

Kubori T and Galán J (2003). Temporal regulation of *Salmonella* virulence factor function by proteasome-dependent protein degradation. *Cell*, **115**, 333–342.

Lindgren S W, Stojilkovic I, and Heffron F (1996). Macrophage killing is an essential virulence mechanism of *Salmonella typhimurium*. *Proc. Natl. Acad. Sci. USA*, **93**, 4197–4201.

Lyczak J B, Cannon C L, and Pier G B (2000). Establishment of *Pseudomonas aeruginosa* infection: Lessons from a versatile opportunist. *Microbes Infect.*, **2**, 1051–1060.

Marcus S L, Wenk M R, Steele-Mortimer O, and Finlay B B (2001). A synaptojanin-homologous region of *Salmonella typhimurium* SigD is essential for inositol phosphatase activity and Akt activation. *FEBS Lett.*, **494**, 201–207.

McDonald C, Vacratsis P O, Bliska J B, and Dixon J E (2003). The *Yersinia* virulence factor YopM forms a novel protein complex with two cellular kinases. *J. Biol. Chem.*, **278**, 18514–18523.

McGuffie E M, Frank D W, Vincent T S, and Olson J C (1998). Modification of Ras in eukaryotic cells by *Pseudomonas aeruginosa* exoenzyme S. *Infect. Immun.*, **66**, 2607–2613.

McNamara B P, Koutsouris A, O'Connell C B, Nougayrede J P, Donnenberg M S, and Hecht G (2001). Translocated EspF protein from enteropathogenic *Escherichia coli* disrupts host intestinal barrier function. *J. Clin. Invest.*, **107**, 621–629.

Menard R, Prevost M C, Gounon P, Sansonetti P, and Dehio C (1996). The secreted Ipa complex of *Shigella flexneri* promotes entry into mammalian cells. *Proc. Natl. Acad. Sci. USA*, **93**, 1254–1258.

Mills S D, Boland A, Sory M P, van der Smissen P, Kerbourch C, Finlay B B, and Cornelis G R (1997). *Yersinia enterocolitica* induces apoptosis in macrophages by a process requiring functional type III secretion and translocation mechanisms and involving YopP, presumably acting as an effector protein. *Proc. Natl. Acad. Sci. USA*, **94**, 12638–12643.

Monack D M, Mecsas J, Ghori N, and Falkow S (1997). *Yersinia* signals macrophages to undergo apoptosis and YopJ is necessary for this cell death. *Proc. Natl. Acad. Sci. USA*, **94**, 10385–10390.

Monack D M, Raupach B, Hromockyj A E, and Falkow S (1996). *Salmonella typhimurium* invasion induces apoptosis in infected macrophages. *Proc. Natl. Acad. Sci. USA*, **94**, 9833–9838.

Murli S, Watson R O, and Galán J E (2001). Role of tyrosine kinases and the tyrosine phosphatase SptP in the interaction of *Salmonella* with host cells. *Cell. Microbiol.*, **3**, 795–810.

Niebuhr K, Jouihri N, Allaoui A, Gounon P, Sansonetti P J, and Parsot C (2000). IpgD, a protein secreted by the type III secretion machinery of *Shigella flexneri*, is chaperoned by IpgE and implicated in entry focus formation. *Mol. Microbiol.*, **38**, 8–19.

Niebuhr K, Giuriato S, Pedron T, Philpott D J, Gaits F, Sable J, Sheetz M P, Parsot C, Sansonetti P J and Payrastre B (2002). Conversion of PtdIns(4,5)P(2) into PtdIns(5)P by the *Shigella flexneri* effector IpgD reorganizes host cell morphology. *EMBO J.*, **21**, 5069–5078.

Norris F A, Wilson M P, Wallis T S, Galyov E E, and Majerus P W (1998). SopB, a protein required for virulence of *Salmonella dublin*, is an inositol phosphate phosphatase. *Proc. Natl. Acad. Sci. USA*, **95**, 14057–14059.

Olson J C, Fraylick J E, McGuffie E M, Dolan K M, Yahr T L, Frank D W, and Vincent T S (1999). Interruption of multiple cellular processes in HT-29 epithelial cells by *Pseudomonas aeruginosa* exoenzyme S. *Infect. Immun.*, **67**, 2847–54.

Orth K (2002). Function of the *Yersinia* effector YopJ. *Curr. Opin. Microbiol.*, **5**, 38–43.

Orth K, Palmer L E, Bao Z Q, Stewart S, Rudolph A E, Bliska J B, and Dixon J E (1999). Inhibition of the mitogen-activated protein kinase kinase superfamily by a *Yersinia* effector. *Science*, **285**, 1920–1923.

Orth K, Xu Z, Mudgett M B, Bao Z Q, Palmer L E, Bliska J B, Mangel W F, Staskawicz B, and Dixon J E (2000). Disruption of signalling by *Yersinia* effector YopJ, a ubiquitin-like protein protease. *Science*, **290**, 1594–1597.

Pederson K J, Vallis A J, Aktories K, Frank D W, and Barbieri J T (1999). The amino-terminal domain of *Pseudomonas aeruginosa* ExoS disrupts actin filaments via small-molecular-weight GTP-binding proteins. *Mol. Microbiol.*, **32**, 393–401.

Persson C, Nordfelth R, Andersson K, Forsberg A, Wolf-Watz H, and Fallman M (1999). Localization of the *Yersinia* PTPase to focal complexes is an important virulence mechanism. *Mol. Microbiol.*, **33**, 828–838.

Phillips R M, Six D A, Dennis E A, and Ghosh P (2003). *In vivo* phospholipase activity of the *Pseudomonas aeruginosa* cytotoxin ExoU and protection of mammalian cells with phospholipase A2 inhibitors. *J. Biol. Chem.*, **278**, 41326–41332.

Riese M J, Wittinghofer A, and Barbieri J T (2001). ADP ribosylation of Arg41 of Rap by ExoS inhibits the ability of Rap to interact with its guanine nucleotide exchange factor, C3G. *Biochemistry*, **40**, 3289–94.

Rosqvist R, Bolin I, and Wolf-Watz H (1988). Inhibition of phagocytosis in *Yersinia pseudotuberculosis*: A virulence plasmid-encoded ability involving the Yop2b protein. *Infect. Immun.*, **56**, 2139–2143.

Rosqvist R, Forsberg A, and Wolf-Watz H (1991). Intracellular targeting of the *Yersinia* YopE cytotoxin in mammalian cells induces actin microfilament disruption. *Infect. Immun.*, **59**, 4562–4569.

Ruckdeschel K, Mannel O, Richter K, Jacobi C A, Trulzsch K, Rouot B, and Heese-mann J (2001). *Yersinia* outer protein P of *Yersinia enterocolitica* simultane-ously blocks the nuclear factor-kappa B pathway and exploits lipopolysac-charide signalling to trigger apoptosis in macrophages. *J. Immunol.*, **166**, 1823–1831.

Ruiz-Albert J, Yu X J, Beuzon C R, Blakey A N, Galyov E E, and Holden D W (2002). Complementary activities of SseJ and SifA regulate dynamics of the *Salmonella typhimurium* vacuolar membrane. *Mol. Microbiol.*, **44**, 645–661.

Salcedo S P and Holden D W (2003). SseG, a virulence protein that targets *Salmonella* to the Golgi network. *EMBO J.*, **22**, 5003–5014.

Sansonetti P J (2001). Rupture, invasion and inflammatory destruction of the intestinal barrier by *Shigella*, making sense of prokaryote-eukaryote cross-talks. *FEMS Microbiol. Rev.*, **25**, 3–14.

Sato H, Frank D W, Hillard C J, Feix J B, Pankhaniya R R, Moriyama K, Finck-Barbancon V, Buchaklian A, Lei M, Long R M, Wiener-Kronish J, and Sawa T (2003). The mechanism of action of the *Pseudomonas aeruginosa*-encoded type III cytotoxin, ExoU. *EMBO J.*, **22**, 2959–2969.

Sauvonnet N, Lambermont I, van der Bruggen P, and Cornelis G R (2002). YopH prevents monocyte chemoattractant protein 1 expression in macrophages and T-cell proliferation through inactivation of the phosphatidylinositol 3-kinase pathway. *Mol. Microbiol.*, **45**, 805–815.

Schesser K, Spiik A K, Dukuzumuremyi J M, Neurath M F, Pettersson S, and Wolf-Watz H (1998). The *yopJ* locus is required for *Yersinia*-mediated inhibition of NF-κB activation and cytokine expression: YopJ contains a eukaryotic SH2-like domain that is essential for its repressive activity. *Mol. Microbiol.*, **28**, 1067–1079.

Shao F, Merritt P M, Bao Z, Innes R W, and Dixon J E (2002). A *Yersinia* effector and a *Pseudomonas* avirulence protein define a family of cysteine proteases functioning in bacterial pathogenesis. *Cell*, **109**, 575–588.

Shao F, Vacratsis P O, Bao Z, Bowers K E, Fierke C A, and Dixon J E (2003). Biochemical chracterization of the *Yersinia* YopT protease: Cleavage site and recognition element in Rho GTPases. *Proc. Natl. Acad. Sci. USA*, **100**, 904–909.

Skoudy A, Nhieu G T, Mantis N, Arpin M, Mounier J, Gounon P, and Sansonetti P (1999). A functional role for ezrin during *Shigella flexneri* entry into epithelial cells. *J. Cell Sci.*, **112**, 2059–68.

Steele-Mortimer O, Knodler L A, Marcus S L, Scheid M P, Goh B, Pfeifer C G, Duronio V, and Finlay B B (2000). Activation of Akt/protein kinase B in epithelial cells by the *Salmonella typhimurium* effector sigD. *J. Biol. Chem.*, **275**, 37718–37724.

Stender S, Friebel A, Linder S, Rohde M, Mirold S, and Hardt W D (2000). Identification of SopE2 from *Salmonella typhimurium*, a conserved guanine nucleotide exchange factor for Cdc42 of the host cell. *Mol. Microbiol.*, **36**, 1206–1221.

Stevens M P, Friebel A, Taylor L A, Wood M W, Brown P J, Hardt W D, and Galyov E E (2003). A *Burkholderia pseudomallei* type III secreted protein, BopE, facilitates bacterial invasion of epithelial cells and exhibits guanine nucelotide exchange factor activity. *J. Bacteriol.*, **185**, 4992–4996.

Sundin C, Henriksson M L, Hallberg B, Forsberg A, and Frithz-Lindsten E (2001). Exoenzyme T of *Pseudomonas aeruginosa* elicits cytotoxicity without interfering with Ras signal transduction. *Cell. Microbiol.*, **3**, 237–246.

Takai Y, Sasaki T, and Matosaki T (2001). Small GTP-binding proteins. *Physiol. Rev.*, **81**, 153–208.

Terebiznik M R, Vieira O V, Marcus S L, Slade A, Yip C M, Trimble W S, Meyer T, Finlay B B, and Grinstein S (2002). Elimination of host cell PtdIns(4,5)P(2) by bacterial SigD promotes membrane fission during invasion by *Salmonella*. *Nat. Cell. Biol.*, **4**, 766–773.

Tran Van Nhieu G, Bourdet-Sicard R, Dumenil G, Blocker A, and Sansonetti P J (2000). Bacterial signals and cell responses during *Shigella* entry into epithelial cells. *Cell. Microbiol.*, **2**, 187–193.

Van Nhieu G T, Caron E, Hall A, and Sansonetti P J (1999). IpaC induces actin polymerization and filopodia formation during *Shigella* entry into epithelial cells. *EMBO J.*, **18**, 3249–3262.

Uchiya K, Tobe T, Komatsu K, Suzuki T, Watarai M, Fukuda I, Yoshikawa M, and Sasakawa C (1995). Identification of a novel virulence gene, *virA*, on the large plasmid of *Shigella*, involved in invasion and intercellular spreading. *Mol. Microbiol.*, **17**, 241–250.

Vallis A J, Finck-Barbancon V, Yahr T L, and Frank D W (1999). Biological effects of *Pseudomonas aeruginosa* type III-secreted proteins on CHO cells. *Infect. Immun.*, **67**, 2040–2044.

van der Velden A W, Lindgren S W, Worley M J, and Heffron F (2000). *Salmonella* pathogenicity island 1-independent induction of apoptosis in infected macrophages by *Salmonella enterica* serotype Typhimurium. *Infect. Immun.*, **68**, 5702–5709.

Vanhaesebroeck B and Alessi D R (2000). The PI3K-PDK1 connection: more than just a road to PKB. *Biochem. J.*, **346**, 561–576.

Viboud G I, So S S K, Ryndak M B, and Bliska J B (2003). Proinflammatory signalling stimulated by the type III translocation factor YopB is counteracted by multiple effectors in epithelial cells infected with *Yersinia pseudotuberculosis*. *Mol. Microbiol.*, **47**, 1305–1315.

Vincent T S, Fraylick J E, McGuffie E M, and Olson J C (1999). ADP-ribosylation of oncogenic Ras proteins by *Pseudomonas aeruginosa* exoenzyme S in vivo. *Mol. Microbiol.*, **32**, 1054–1064.

Visser L G, Annema A, and vanFurth R (1995). Role of Yops in inhibition of phagocytosis and killing of opsonized *Yersinia enterocolitica* by human granulocytes. *Infect. Immun.*, **63**, 2570–2575.

von Pawel-Rammingen U, Telepnev M V, Schmidt G, Aktories K, Wolf-Watz H, and Rosqvist R (2000). GAP activity of the *Yersinia* YopE cytotoxin specifically targets the Rho pathway: A mechanism for disruption of actin microfilament structure. *Mol. Microbiol.*, **36**, 737–748.

Yao T, Mecsas J, Healy J I, Falkow S, and Chien Y (1999). Suppression of T and B lymphocyte activation by a *Yersinia pseudotuberculosis* virulence factor, YopH. *J. Exp. Med.*, **190**, 1343–50.

Waterman S R and Holden D W (2003). Functions and effectors of the *Salmonella* pathogenicity island 2 type III secretion system. *Cell. Microbiol.*, **5**, 510–511.

Yoshida S, Katayama E, Kuwae A, Mimuro H, Suzuki T, and Sasakawa C (2002). *Shigella* deliver an effector protein to trigger host microtubule destabilization, which promotes Rac1 activity and efficient bacterial internalization. *EMBO J.*, **21**, 2923–2935.

Zhou D, Chen L M, Hernandez L, Shears S B, and Gálan J E (2001). A *Salmonella* inositol polyphosphatase acts in conjunction with other bacterial effectors to promote host cell actin cytoskeleton rearrangements and bacterial internalization. *Mol. Microbiol.*, **39**, 248–259.

Zumbihl R, Aepfelbacher M, Andor A, Jacobi C A, Ruckdeschel K, Rouot B, and Heesemann J (1999). The cytotoxin YopT of *Yersinia enterocolitica* induces modification and cellular redistribution of the small GTP-binding protein RhoA. *J. Biol. Chem.* **274**, 29289–29293.

Zychlinsky A, Kenny B, Menard R, Prevost M C, Holland I B, and Sansonetti P J (1994). IpaB mediates macrophage apoptosis induced by *Shigella flexneri*. *Mol. Microbiol.*, **11**, 619–627.

Zychlinsky A, Prevost M C, and Sansonetti P J (1992). *Shigella flexneri* induces apoptosis in infected macrophages. *Nature*, **358**, 167–169.

Zychlinsky A, Thirumalai K, Arondel J, Cantey J R, Aliprantis A O, and Sansonetti P J (1996). In vivo apoptosis in *Shigella flexneri* infections. *Infect. Immun.* **64**, 5357–5365.

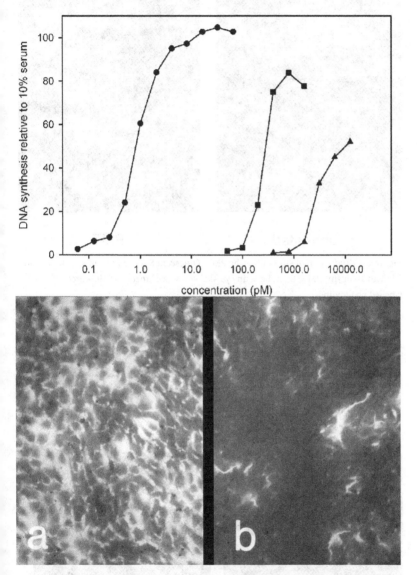

Figure 2.1. PMT is a mitogen for Swiss 3T3 cells. Top panel: relative mitogenicity of PMT
(•), platelet derived growth factor (PDGF) (■) and bombesin (▲); lower panel: cell
proliferation induced by 48 h PMT treatment: a, untreated cells; b, PMT treated cells.

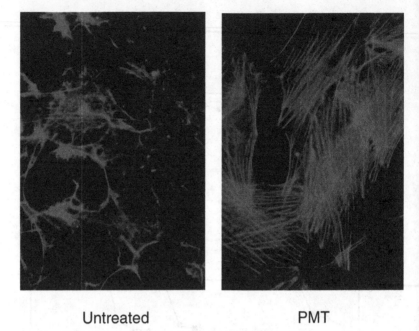

Untreated   PMT

Figure 2.3. Induction of stress fibres by PMT. Quiescent Swiss 3T3 cells were treated for 8 h with 20 ng/ml PMT, then the actin cytoskeleton was stained with fluorescently labelled phalloidin.

1   1285

R   T   C

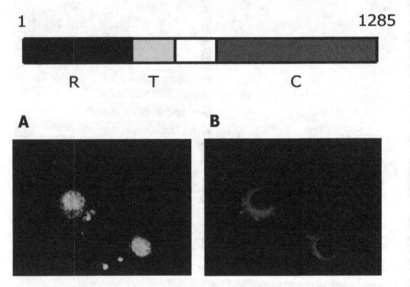

Figure 2.4. The functional domains of PMT. Top panel: diagram showing the approximate locations of the functional domains. R, receptor-binding domain; T, membrane translocation domain; C, catalytic domain. Lower panel: quiescent Swiss 3T3 cells were microinjected with the C-terminal of PMT (residues 681–1285) and with rabbit IgG. DNA synthesis was assayed by addition of BrdU, which is incorporated into the DNA of activated cells. A, green nuclei represent BrdU positive cells; B, microinjected cells stained red.

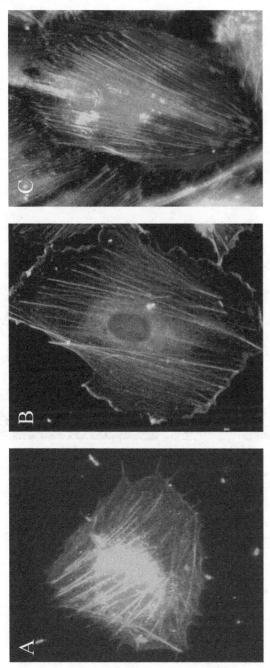

Figure 5.2. *B. bacilliformis* infected endothelial cells showing the structures associated with the activation of Rho-GTPases. A. Infected endothelial cell stained with F-actin (green) and anti-Cdc42 antibodies (red) showing microspikes related to activation of Cdc42 after 30 min of infection. B. Infected endothelial cell stained with F-actin (green) and anti-Rac antibodies (red) showing membrane ruffling and colocalization of F-actin and Rac in the membrane ruffles (yellow) related to activation of Rac after 1 h of infection. C. Infected endothelial cell stained with F-actin (green) and anti-paxillin antibodies (red) showing formation of thick stress fibers terminating in paxillin-rich focal adhesions (red) after 24 hrs of infection with the activation of Rho.

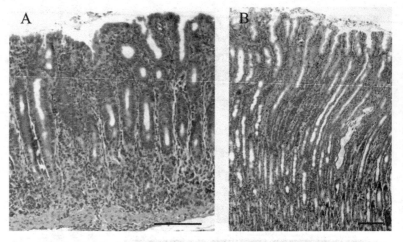

Figure 8.2. Gastric histology in Mongolian gerbil infected with *H. pylori* strain SS1 for 36 weeks. Haematoxylin and Eosin stained sections of A) antral mucosa and B) corpus mucosa. Bar = 100 μm. Adapted from Naumann and Crabtree (2004).

CHAPTER 7

Bacterial toxins and bone remodelling

Neil W A McGowan, Dympna Harmey, Fraser P Coxon,
Gudrun Stenbeck, Michael J Rogers, and Agamemnon E Grigoriadis

Bacterial protein toxins are powerful biological poisons normally associated with impairment of cellular function and/or cellular death. The wide spectrum of physiological processes and cell types that are affected by bacterial products also includes bone tissue and bone cells. It has been known for many years that bacterial infection or exposure to certain toxins can lead to pathological bone disorders, most commonly, those associated with abnormal or excessive bone loss, such as periodontal disease (reviewed by Henderson and Nair, 2003). However, in most cases the bone-resorbing factors involved in these effects remain part of, or associated with, the bacterial surface. For example, the bone-resorbing effects of endotoxin, a component of lipopolysaccharide, are well established, although for the most part this action appears to be indirect, being dependent on the production of pro-inflammatory cytokines (IL-1, TNFα) from other cell types (Nair et al., 1996; Henderson and Nair 2003). In contrast, the effects of bacterial protein toxins on the cellular constituents of bone remain largely unknown. For simplicity, this review will focus only on bacterial toxins, in particular, those toxins that interfere with key signalling processes that have direct relevance to bone cell differentiation and function. However, a brief overview of the general biology of bone cells is necessary before discussing the mechanisms of toxin action and specific signal transduction pathways in bone.

BONE

Throughout life the vertebrate skeleton is in a constant state of turnover. Physiological bone remodelling requires the tight coupling of bone degradation to bone formation, in order for the precise replacement of old or damaged bone

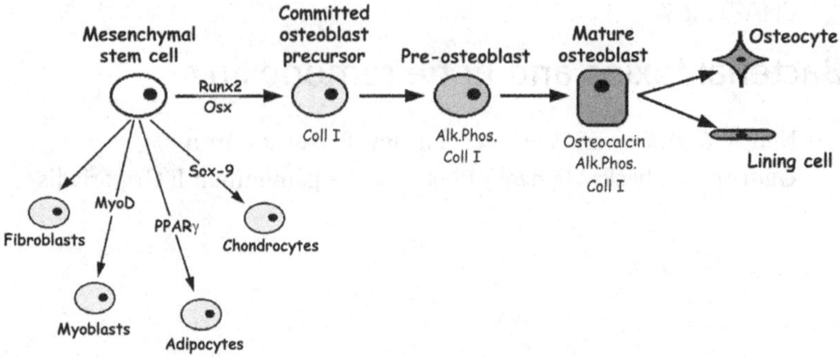

Figure 7.1. The osteoblast lineage. Osteoblasts are derived from mesenchymal stem cells, which have the potential to differentiate into other mesenchymal derivatives such as muscle, fat, cartilage, and other connective tissue cells. The commitment to each lineage is dependent upon lineage-specific transcription factors, such as MyoD for muscle, PPARγ for adipocytes, Sox-9 for chondrocytes, and runx2/cbfa1 and Osterix (Osx) for osteoblasts. The expression of genes such as type I collagen (coll I), alkaline phosphatase (Alk.Phos.), and osteocalcin are commonly used as markers for the osteoblast lineage, although osteocalcin is the only osteoblast-specific gene (see Aubin, 1998, for details).

to occur, and for maintenance of bone mass and skeletal homeostasis. Disruption of this balance leads to an uncoupling of remodelling and ultimately skeletal abnormalities characterised either by excessive bone loss (e.g., osteoporosis) or, alternatively, by increases in bone mass (e.g., osteopetrosis). Bone remodelling is controlled by the concerted actions of essentially two cell types: osteoblasts, the cells that form bone, and osteoclasts, the cells that resorb or degrade bone.

Osteoblast Differentiation and Bone Formation

Osteoblasts are post-mitotic cells that exhibit a cuboidal morphology and are situated on active bone-forming surfaces. Their basic function is to secrete the organic matrix of bone, largely containing type I collagen, which then proceeds to mineralisation. Osteoblasts are derived from pluripotent mesenchymal stem cells in a differentiation sequence that is a linear process, requiring the sequential activation and suppression of specific genes. They are related at the precursor cell level to other mesenchymal cell derivatives, such as muscle cells, adipocytes, chondrocytes, and other connective tissue fibroblasts (Figure 7.1; see also Aubin, 1998). The identification of osteoblasts *in situ* is aided by morphological criteria, but more importantly by marker genes that are expressed at different stages of differentiation. Recent gene knockout studies in mice have identified two transcription factors, runx2/cbfa-1 and osterix, that are essential for the commitment of cells to the osteoblast lineage and for differentiation to mature, bone-forming osteoblasts (reviewed by Ducy et al., 2000; Karsenty and Wagner, 2002). Besides runx2/cbfa-1 and osterix, the expression of several marker genes is commonly used to indicate cells of the osteoblast lineage, namely, type I collagen, alkaline phosphatase, and osteocalcin, the last being specific to osteoblasts (see also Aubin, 1998; Karsenty and Wagner, 2002, for further details).

The study of the mechanisms controlling bone formation has been aided greatly by the establishment of efficient *in vitro* systems of osteoblast differentiation and bone formation. Primary cultures of osteoblasts, derived either from perinatal rodent calvariae or from adult bone marrow stroma, contain osteoprogenitor cells that can undergo the full differentiation cascade from committed precursors cells to fully functional osteoblasts that are capable of forming three-dimensional, mineralised bone nodules. Such systems have been essential for determining the effects of specific growth factors and transcription factors on osteogenesis using gain- and loss-of-function systems, and are ideal for screening compounds that are thought to regulate bone formation.

Osteoclast Differentiation

Osteoclasts are large multinucleated cells whose function is to resorb or degrade bone. In contrast to osteoblasts, osteoclasts are of haematopoietic origin, more specifically being derived from cells of the monocyte/macrophage lineage (Figure 7.2). However, it was first hypothesised over 20 years ago, and subsequently proven, that the *in vitro* differentiation of osteoclasts from bone marrow cells required the presence of an osteoblastic/stromal component that was the primary target of various osteoclast inductive factors. Moreover, this effect was critically dependent on physical contact between the two cell types, indicating that an inducible, membrane-bound molecule was involved (Rodan and Martin, 1981; Takahashi et al., 1988; Suda et al., 1992).

Recently, the molecules involved in the physiological control of osteoclast formation were identified by two independent groups (for review see Suda et al., 1999). Importantly, these two molecules, receptor activator of NF-κB ligand (RANKL) and macrophage colony-stimulating factor (MCSF), are expressed by osteoblastic/stromal cells, and are both necessary and sufficient for osteoclast formation *in vitro* and *in vivo* (Figure 7.2) (Hofbauer and Heufelder, 1998; Lacey et al., 1998; Yasuda et al., 1998). RANKL binds to its receptor RANK, which is expressed by marrow-derived osteoclast precursors,

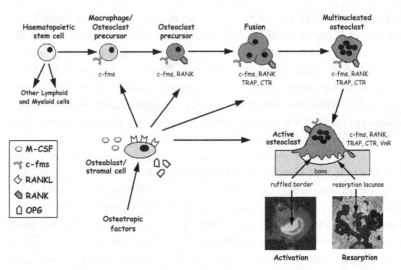

Figure 7.2. The osteoclast lineage and regulation of differentation and activation. Osteoclasts are derived from haematopoietic stem cells, specifically from circulating monocyte/macrophage precursors. The MCSF/c-*fms* signalling pathway provides essential proliferative and survival signals for both macrophages and osteoclasts at all stages of differentiation. The RANKL/RANK signalling pathway is essential for the differentiation, fusion, and survival of osteoclasts, and is inhibited by the soluble decoy receptor, OPG. Osteoclasts can be identified by the expression of specific markers, such as TRAP, CTR, and the $\alpha v \beta 3$ integrin, vitronectin receptor (VnR). Many osteotropic factors, such as vitamin D$_3$ and parathyroid hormone, regulate osteoclast formation and/or activity indirectly by influencing the expression ratio of RANKL:OPG on osteoblastic/stromal cells (see text for details). The micrographs show F-actin staining of the ruffled border in active osteoclasts, and resorption lacunae formed on a dentine substrate.

peripheral blood monocytes, and by mature osteoclasts *in vivo*, and leads to activation of several intracellular signalling cascades involving TNF-receptor-associated factor (TRAF) family members, JNKs, *c-src*, and the serine/threonine kinase Akt/PKB, leading to activation of the transcription factors NF-κB and *c-fos* (Anderson et al., 1997; Darnay et al., 1998; Darnay et al., 1999; Hsu et al., 1999; Wei et al., 2001) (see also Karsenty and Wagner, 2002, for a review). MCSF is also essential for osteoclast differentiation, providing the proliferative signal for osteoclast precursors prior to fusion. MCSF binds to the tyrosine kinase receptor, the proto-oncogene *c-fms*, that is present on osteoclast precursors (Figure 7.2). MCSF/c-*fms* signalling is the primary determinant of the total osteoclast precursor pool and is essential for both their proliferation and differentiation (Sarma and Flanagan, 1996; Suda et al.,

1999; Fujikawa et al., 2001). RANKL serves to commit the pool of MCSF-expanded precursors to the osteoclast phenotype, essentially being involved in all post-proliferative stages of osteoclast ontogeny, such as differentiation, fusion, activation, and survival.

In vivo, the physiological effects of RANKL are negatively regulated by the expression of a soluble decoy receptor termed osteoprotegerin (OPG), which is also synthesised by osteoblasts and stromal cells, and which binds to and sequesters RANKL, preventing it from binding to RANK (Figure 7.2) (Simonet et al., 1997). Thus, the relative ratio of RANKL to OPG is the most important determinant of osteoclast formation *in vivo* and many osteotropic factors, which induce osteoclast formation do so via the osteoblastic/stromal component (Figure 7.2). Osteoblasts and stromal cells are therefore responsible for the production of all three principal factors essential for osteoclastogenesis and thus critically control the whole process. The convergence of multiple signals in the control of osteoclastogenesis by the production of stimulatory and inhibitory molecules represents an extremely efficient and intricate means of regulation.

Mechanisms of Bone Resorption

The resorption of mineralised matrix is an attribute specific to osteoclasts, during which the highly specialised features of these cells become apparent. Osteoclasts express high levels of tartrate resistant acid phosphatase (TRAP) and calcitonin receptors (CTR), but the most dramatic feature is the appearance during resorption of the ruffled border – a complex membranous structure composed of folds and invaginations. The initial resorptive event, essential for the subsequent polarisation of osteoclasts, is the attachment of cells to the surface of bone. Osteoclast adherence and motility critically depend on the expression of $\alpha v \beta 3$ integrin, or vitronectin receptor (VnR), which mediates adherence to a wide range of extracellular matrix proteins (such as fibronectin, vitronectin, osteopontin, and bone sialoprotein) known to be expressed in bone and bone marrow through the recognition of an RGD peptide (Helfrich and Horton, 1999). Adherence to bone results in the generation of an isolated extracellular environment between the osteoclast and bone surface. The segregation of this proteolytic environment is maintained by the clear zone, rich in filamentous actin (F-Actin) and arranged in a ring-like structure surrounding the ruffled border (Vaananen and Horton, 1995).

Osteoclast polarisation involves a variety of signalling proteins, in particular p60[c-src], the importance of which was demonstrated by the osteopetrotic phenotype observed in p60[c-src] knockout mice, resulting in dysfunctional osteoclasts (see below) (Soriano et al., 1991). These mice have multinucleated osteoclasts that attach to bone via the clear zone but fail to form ruffled borders (Boyce et al., 1992; Lowe et al., 1993). Subsequently, a variety of signalling proteins downstream of p60$^{c\text{-}src}$ have been identified (c-Cbl, p130^{cas}, PYK2, Rho p21, and PI-3K), although the precise interplay and hierarchy involved remain unclear (reviewed by Duong et al., 2000).

The acidic and proteolytic microenvironment below the ruffled border favours the demineralisation and degradation of bone matrix. Acidification of resorption lacunae involves the active transport of protons across the ruffled border by a vacuolar H$^+$-ATPase (Blair et al., 1989), leading to pH values in the range of 3–4. The acidic microenvironment aids the dissolution of bone mineral, which is closely followed by the degradation of bone matrix by a variety of lysosomal enzymes secreted by the osteoclast. High concentrations of these enzymes are maintained in the resorption lacunae, and products of bone degradation are subsequently endocytosed by the osteoclast and released at the basolateral membrane (Nesbitt and Horton, 1997; Salo et al., 1997).

The resorption of bone is therefore a multi-step process that involves the migration, proliferation, and commitment of immature osteoclast precursors, formation of multinucleated cells by fusion, and progression from a resting to a resorbing stage by attachment and polarisation on the bone surface, followed by the degradation of both organic and inorganic matrix (Athanasou, 1996; Suda et al., 1997; Teitelbaum, 2000). The importance of each of these steps, and more specifically, the signalling molecules and transcription factors that regulate each step of osteoclast differentiation and function, have been demonstrated unequivocally by gene knockout studies in mice, with genes such as c-*src*, TRAF-6, NF-κB, or *c-fos* to mention but a few (Grigoriadis et al., 1994; Iotsova et al., 1997; Lomaga et al., 1999). Indeed, disruption of any of these genes leads to osteopetrosis, a family of diseases characterised by osteoclast dysfunction. Studies in osteopetrotic mouse mutants have identified the essential factors involved in all stages of osteoclast differentiation, and these are summarised in detail elsewhere (Karsenty and Wagner, 2002). Finally, because each of the steps in osteoclastogenesis involves in some way a tight control and regulation of the actin cytoskeleton, osteoclasts are attractive targets for analysing the role of bacterial toxins that target and perturb signal transduction pathways affecting the cytoskeleton (see below).

Bone Remodelling

The process of bone remodelling maintains the mechanical integrity of the skeleton (Mundy, 1998; Raisz, 1999; Manolagas, 2000). Bone remodelling is a focal process of renewal and repair that occurs throughout life to prevent

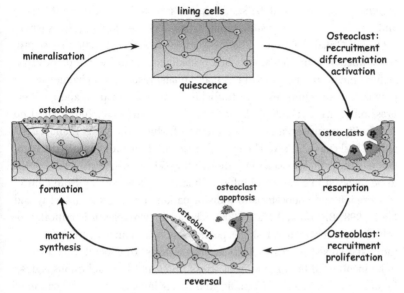

Figure 7.3. The bone remodelling cycle. Bone remodelling occurs in "basic multicellular units" (BMU) and is essential for the maintenance of skeletal integrity. Remodelling can be divided into four phases: resorption, reversal, formation, and quiescence. Remodelling begins with osteoclast precursor recruitment, differentiation, fusion, and activation. Following resorption and osteoclast apoptotic cell death, there is a reversal period in which osteoblast recruitment and proliferation occurs. Osteoblasts lay down new bone in the form of osteoid, which is subsequently mineralised and becomes covered in bone lining cells on reaching a quiescent state. For simplicity, each process is segregated in this diagram. Within every BMU, resorption and formation occur concurrently although in different skeletal locations, and this entire process is tightly controlled by the concerted actions of systemic hormones, local growth factors, and transcriptional regulators.

accumulation of old or damaged bone. At any one time approximately 10% of the adult skeleton is undergoing active remodelling, whereas the remaining 90% is quiescent. All bone remodelling occurs within discrete anatomical units (basic multicellular unit – BMU) where bone resorption by osteoclasts is tightly coupled to bone formation by osteoblasts. The spatial and temporal arrangement of osteoblasts and osteoclasts within a BMU ensures that remodelling maintains a specific cycle that can be divided into four phases: resorption, reversal, formation, and quiescence (Figure 7.3).

BACTERIAL TOXINS, RHO FAMILY GTPases, AND BONE CELLS

Several bacterial toxins target eukaryotic cells, specifically interfering with the normal function of Rho GTPases (see also Chapters 3 and 6). Rho proteins, together with the Ras, Rab, Ran, and Arf subfamilies, make up the large and highly conserved Ras superfamily of small GTP-binding proteins (small GTPases) (Takai et al., 2001; Etienne-Manneville and Hall, 2002). These proteins function as molecular switches, transducing modifications of the extracellular environment into intracellular signals, resulting in changes in cell growth, cytoskeleton, vesicular trafficking, nuclear transport, and gene expression (Takai et al., 2001; Etienne-Manneville and Hall, 2002). The Rho subfamily comprises at least 15 proteins, of which RhoA, Rac1, and Cdc42 are the best characterised. The main function of these proteins is in the regulation of the actin cytoskeleton, although Rho GTPases have also been shown to be involved in a variety of other cellular functions ranging from control of secretion and endocytosis to transformation and apoptosis (Mackay and Hall, 1998; Bishop and Hall, 2000). The actin cytoskeleton is critically involved in the control of cell shape, polarity, and adhesion in addition to other, more specialised processes such as phagocytosis. In general, Rho appears to be involved in the formation of stress fibres and focal adhesions, Cdc42 induces the formation of filopodia, and Rac is involved in the formation of lamellipodia and membrane ruffles (Nobes and Hall, 1995).

Bacterial toxins have proved extremely useful in the elucidation of several signalling pathways utilised by Rho GTPases (reviewed in Chapter 3; see also Lerm et al., 2000). Covalent modification of Rho GTPases has been shown to disrupt normal actin cytoskeletal regulation with varying degrees of specificity. Toxins that act in this fashion include *Clostridium botulinum* C3 ADP-ribosyltransferase, the related C3-like exoenzymes, and the clostridial cytotoxins, *Clostridium difficile* toxins A and B. *Clostridium botulinum* C3 exoenzyme (C3 transferase) irreversibly ADP-ribosylates Rho A, B, and C at Asn[41] resulting in a block in Rho-mediated signalling, whereas the less specific, more promiscuous *Clostridium difficile* toxins A and B inactivate all Rho GTPase family members (Rho, Rac, and Cdc42) by glucosylating the nucleotide-binding site. A similar mechanism is involved in the action of the lethal and haemorrhagic toxins from *Clostridium sordellii* and the α-toxin from *Clostridium novyi*.

Conversely, bacterial toxins have also been shown to activate the small GTPases by deamidation or transglutamination. Cytotoxic necrotizing factor (CNF) from *Escherichia coli* and dermonecrotic toxin (DNT) from *Bordetella* species catalyse the deamidation of Rho at Gln[63] (Flatau et al., 1997; Schmidt et al., 1997; Lerm et al., 1999). The enzyme activity of CNF removes the carboxamide nitrogen of Gln[63], which is required for the correct positioning of GTP for hydrolysis in the catalytic pocket of the GTPases. Thus, CNF inhibits the intrinsic and GAP-stimulated GTPase activities, thereby resulting in the constitutive activation of Rho. The catalytic domain of CNF is located at the carboxy-terminal of the peptide, and Cys^{866} and His^{881} histidine are implicated in its deamidase activity (Schmidt et al., 1998). The action of CNF is not confined to Rho; it also targets the other members of the Rho GTPases like Rac and Cdc42 (see also Boquet (2000) for review).

When Rho from DNT treated cells was analysed it was found that DNT catalyses a different modification than that induced by CNF. DNT possesses transglutaminase activity, thereby attaching primary amines onto Rho at position Gln^{63}. Further comparison of the enzyme activities of both CNF and DNT have found that both toxins are capable of catalysing deamidation and transglutamination reactions, but DNT preferentially acts as a transglutaminase (Horiguchi et al., 1997; Schmidt et al., 1998; Schmidt et al., 1999).

With respect to bone, DNT has been shown to cause bone loss when injected over rodent calvariae, or in *in vitro* assays, although the mechanisms underlying these effects are not known (Kimman et al., 1987; Horiguchi et al., 1995). Thus, elaboration of the actions of these toxins has demonstrated their specific and selective targeting of the Rho GTPases and has made them useful tools in investigating the complex signalling pathways mediated via the small GTPases, in particular, in bone tissue.

Pasteurella multocida Toxin and Bone

Pasteurella multocida toxin (PMT) is another bacterial toxin that provides an ideal tool for the study of bone cell signalling *in vitro*, particularly as this toxin has profound effects on bone physiology *in vivo*. PMT is a large, 146kDa intracellularly acting protein that is the most potent mitogen known for fibroblasts, stimulating DNA synthesis at low picomolar concentrations (Rozengurt et al., 1990). A detailed description of the biochemistry and molecular biology of this toxin can be found in Chapter 2. However, of relevance to bone physiology, PMT is an extremely interesting toxin, as it is associated with a disease in pigs called atrophic rhinitis, which is characterised by the progressive loss of the nasal turbinate bones (Ackermann et al., 1991; Lax and Chanter, 1990). That PMT is the principle causative agent of atrophic rhinitis was shown unequivocally by Lax and co-workers, who demonstrated that injection of recombinant PMT into piglets recapitulated the bone disease, whereas an inactive mutant, C1165S, with identical biochemical properties to the wild-type recombinant protein but with no biological activity (see Chapter 2), had no effect (Ward et al., 1998). How PMT causes atrophic rhinitis is not well understood (see below), although it is very likely that its intracellular target(s) play(s) a role in altering the behaviour of bone cell populations.

Several signalling cascades are affected by PMT, both directly and indirectly (reviewed by Ward et al., 1998) (see also Chapter 2). The toxin stimulates signalling cascades linked to phospholipase Cβ, leading to increased formation of diacylglycerol and inositol phosphates, increased protein kinase C activity, and ultimately mobilisation of intracellular calcium. Interestingly with regard to osteoclast function, PMT also stimulates signalling pathways linked to the cytoskeleton, via a mechanism that involves Rho GTPase. Inhibitors of Rho and downstream effectors block the cytoskeletal rearrangements induced by PMT (Lacerda et al., 1997), although Rho itself does not appear to be a direct target for this toxin. Despite the molecular target for PMT being unknown, recent studies in transgenic mice have provided strong evidence to suggest that activation by PMT requires a G_q specific event (Zywietz et al., 2001). Furthermore, cAMP levels are not affected by PMT, suggesting that unlike cholera toxin (from *Vibrio cholerae*) and pertussis toxin (from *Bordetella pertussis*), PMT does not target the G proteins, G_i and or G_s (Rozengurt et al., 1990).

Thus, with regard to the bone pathology observed in pigs, it is not yet clear whether this toxin stimulates processes resulting in excessive bone resorption, or alternatively, defective bone formation because either or both could explain the observed phenotype. Previous studies have utilised *in vitro* and *in vivo* bone cell and organ culture systems to address these questions, although the findings are not clear and often contradictory, reflecting the diversity and heterogeneity of the particular culture systems employed (see below). The challenge therefore is to identify the precise cellular targets (i.e., osteoblasts vs osteoclasts) responding to PMT, and to delineate which of the proposed signal transduction pathways activated by PMT control the differentiation and activity of these bone cell populations.

PMT and Osteoblasts

As mentioned above, PMT is a potent mitogen for fibroblasts (Rozengurt et al., 1990) and it is possible that the proliferation of other mesenchymal cell derivatives may also be stimulated by PMT. Analysis of PMT effects of cell proliferation and DNA synthesis in different osteoblast populations, such as primary rat and chick osteoblasts and osteoblast-like osteosarcoma cells, demonstrated that PMT is clearly mitogenic for osteoblasts (Figure 7.4a; see also Harmey et al., 2004). Moreover, *in vitro* bone formation assays, whereby primary osteoblast precursors are stimulated to undergo *in vitro* differentiation into functional osteoblasts and lay down bone matrix in three-dimensional nodules, demonstrated that PMT was a potent inhibitor

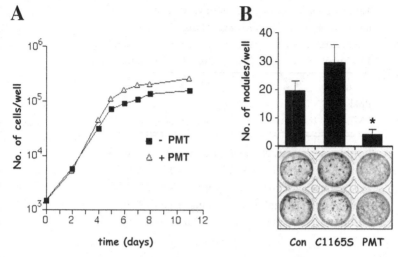

A

B

Figure 7.4. PMT regulates osteoblast proliferation and differentiation. (A) Effects of PMT on the proliferation of primary rat calvarial osteoblasts. Cells were plated in the absence or presence of recombinant PMT (20 ng/ml) and cell number was quantified on the indicated days. (B) Effects of PMT on *in vitro* differentiation of rat calvarial osteoblasts into bone nodules. Cells were cultured for 15 days in the absence (Con) or presence of recombinant PMT (20 ng/ml) or inactive mutant PMT C1165S (20 ng/ml), and mineralised bone nodules were quantified (upper panel) and visualised using the von Kossa technique (black staining – lower panel). ($*p < 0.05$ vs control) (see also Harmey et al., 2004).

of osteoblast differentiation and bone nodule formation (Figure 7.4B; see also Harmey et al., 2004). The inhibitory effect of PMT was also confirmed by the downregulation of the expression and activity of several osteoblast markers (e.g., osteonectin, type I collagen, alkaline phosphatase, osteocalcin) (Sterner-Kock et al., 1995; Mullan and Lax, 1996; Harmey et al., 2004). We have also recently demonstrated that PMT downregulates the expression of the runx2/cbfa-1 transcription factor, which, as shown in Figure 7.1, is an essential gene for osteoblast differentiation and bone formation (Harmey et al., 2004). The mechanism of PMT inhibition of osteoblast differentiation is not well understood, although the signalling pathways known to be triggered by PMT might offer some clue as to its mechanism of action in osteoblasts. Indeed, PMT treatment of osteoblasts induced cytoskeletal rearrangements, suggesting activation of the Rho GTPase, and our recent data have demonstrated that the PMT-induced inhibition of osteoblast differentiation was restored by treatment with the Rho inhibitor, C3 transferase (Table 7.1 and Harmey et al., 2004).

Table 7.1 *The effects of PMT on* in vitro *bone nodule formation of primary osteoblasts*

Treatment	No. of bone nodules
Control	44 ± 4
PMT	2 ± 0.5
PMT + C3 transferase	50 ± 12.5

Primary mouse osteoblasts were treated in the absence (Control) or presence of recombinant PMT (20 ng/ml) and the number of bone nodules were assessed. Treatment with the Rho inhibitor C3 transferase abolishes the inhibitory effects of PMT. ($*p < 0.05$ vs control).

Taken together, these findings have demonstrated that activation of the Rho GTPase negatively regulates osteoblast differentiation, and have also demonstrated the usefulness of PMT in identifying a signalling pathway in osteoblasts that is important for their function in synthesising bone.

PMT and Osteoclasts

PMT causes bone loss *in vivo*, and the results described in the previous section suggest that one mechanism that may contribute to this phenotype is the ability of PMT to inhibit bone formation in the bone remodelling cascade. However, the progressive and rapid loss of nasal turbinate bones in the pig is well established, and points to some effects of PMT on bone-resorbing osteoclasts. Bone remains a principal target for PMT, an observation that occurs apparently regardless of the route of administration. The effects of PMT, however, are not restricted to pigs, or indeed to the nasal turbinate bones, as pathological resorption has also been detected in rats and the long bones of foetal mice (Kimman et al., 1987; Ackermann et al., 1991; Felix et al., 1992; Martineau-Doize et al., 1993). The pro-resorptive effects of PMT are therefore well established, but the effects of PMT on the osteoclast itself remain unclear. Despite the slight increases *in vivo* in osteoclast size and number observed by some (Martineau-Doize et al., 1993), findings to the contrary have also been reported (Ackermann et al., 1993). Moreover, *in vitro* studies have been unable to elaborate further on the direct effects of PMT on osteoclast differentiation and activity, relative to the bone loss observed *in vivo*.

As described earlier, osteoblasts are essential in osteoclastogenesis, and any pathological alteration in bone mass could be attributed to effects on osteoblasts, osteoclasts, or both. Long-term porcine and murine bone marrow cultures in the presence of PMT have shown increases in osteoclast-like cell formation (TRAP positive multinucleated cells) (Jutras and Martineau-Doize, 1996; Gwaltney et al., 1997). However, in the studies by Gwaltney et al. (1997), no increase in resorption was observed relative to the increase in osteoclast-like cell number. Moreover, resorption data were not included in the findings reported by Jutras and Martineau-Doize (1996). The only truly defining characteristic of *bona fide* osteoclasts requires functional criteria, namely, the ability to form resorption lacunae on a mineralised substrate (i.e., bone or dentine). In this regard, it is imperative that osteoclast formation assays incorporate a functional element. Furthermore, the heterogeneous nature of bone marrow cultures and high probability of stromal cell contamination in these culture systems could be contributing indirectly to the pro-resorptive effects.

It is interesting to note therefore that in the studies by Gwaltney et al. (1997) it was suggested that marrow stromal cells may be involved in the observed increase in osteoclast formation. Subsequent studies by Mullan and Lax, using partly purified populations of chick osteoblasts and osteoclasts, found no effect of PMT on bone resorption when either cell population was cultured separately (Mullan and Lax, 1998). However, when co-cultured with cell contact permitted, bone resorption was stimulated by PMT in a concentration-dependent manner, indicating that the bone-resorbing effects of PMT appear to be mediated, at least in part, via an interaction with osteoblasts (see below) (Mullan and Lax, 1998). This suggested that PMT induced a pro-resorptive signal via osteoblasts expressed on the membrane of these cells. By analogy with a variety of other mediators that act via the osteoblast to induce bone resorption, this would suggest that PMT may increase the expression ratio of RANKL to OPG, although our studies in murine osteoblasts suggest that the ratio of RANKL relative to OPG is decreased by PMT, indicating an overall negative effect on osteoclastogenesis (Harmey and Grigoriadis, unpublished).

Following the discovery of the RANKL/OPG system of soluble regulators of osteoclastogenesis, it has been possible to investigate osteoclast differentiation much more efficiently and unambiguously in purified populations of osteoclast precursors – for example, using macrophage precursors from murine bone marrow preparations or human monocytes from peripheral blood. Recently our group has investigated the role of PMT on osteoclast differentiation and activity using this system, which has so far not been

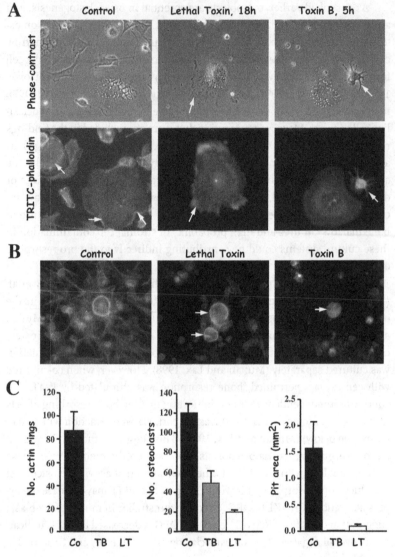

Figure 7.5. Toxin B and Lethal Toxin interfere with osteoclast morphology and function. (A) Images of rabbit osteoclasts on plastic dishes were taken at the indicated time periods, and cells were then fixed and stained with TRITC-phalloidin. Both lethal toxin and toxin B disrupted the distribution of peripheral actin (arrows in Control) and caused retraction of osteoclasts (retraction fibres indicated by arrows). Longer incubation with both toxins resulted in complete retraction and loss of adherence (not shown). (B) Rabbit osteoclasts were treated for 20 hours on dentine slices, then fixed and stained with TRITC-phalloidin. Both lethal toxin and toxin B caused disruption of F-Actin rings and retraction of

addressed by others. Following the demonstration of the potent inhibitory effect of PMT on osteoblast differentiation – an effect partially mediated by activation of Rho – we found that PMT exerted a potent inhibitory effect on murine osteoclastogenesis and resorption directly via acting on the osteoclast precursor population (Harmey and Grigoriadis, unpublished). Furthermore, PMT also inhibited human osteoclast formation from peripheral blood-derived osteoclast precursors (McGowan and Grigoriadis, unpublished), and in both precursor populations, these direct inhibitory actions of PMT were restricted to the early stages of culture. Interestingly, preliminary data also suggest that PMT may elicit an activating effect in mature human osteoclasts – an effect also associated with morphological changes in the F-Actin rings expressed by these cells (McGowan and Grigoriadis, unpublished). These results suggest that PMT may have significantly different effects on the differentiation of osteoclasts from haematopoietic precursors, versus its effects on mature osteoclasts. It remains to be seen which signalling mechanisms activated by PMT are involved in these divergent effects.

GTPases AND OSTEOCLASTS

With the important role of Rho GTPases in the physiological regulation of the actin cytoskeleton, and the requirement for unique cytoskeletal arrangements in normal osteoclast function (polarisation, ruffled border formation, and subsequent resorption), it would be predicted that bacterial toxins might influence osteoclast morphology and activity. Indeed, treatment of primary mature osteoclasts with either *Clostridium difficile* toxin B or lethal toxin from *Clostridium sordellii*, which inactivate all Rho family members (Rho, Rac, and Cdc42), disrupted normal F-Actin ring formation and inhibited bone resorption (Figure 7.5; Coxon and Rogers, unpublished observations). These data are in support of previous studies by Zhang and colleagues, who, using C3 exoenzyme, were the first to suggest a dependence on Rho for normal F-Actin ring (sealing zone) formation and resorption *in vitro* (Zhang et al., 1995). RhoA is highly expressed in osteoclasts and appears to be actively involved in the formation of podosomes – punctate membrane protrusions that are rich in F-Actin and proteins such as vinculin and talin. Chellaiah and colleagues,

---

osteoclasts (arrows). (C) Rabbit osteoclasts on dentine were treated for 48 hours, and the number of F-Actin rings, osteoclasts, and resorption area were determined. Both toxin B (TB) and lethal toxin (LT) completely disrupted actin rings, reduced the number of adherent osteoclasts on dentine, and dramatically inhibited bone resorption.

using C3 exoenzyme in addition to constitutively active (Rho^{Val-14}) and dominant negative forms of Rho (Rho^{Asn-19}), demonstrated that Rho activation stimulated podosome assembly, osteoclast motility, and the resorptive activity of avian osteoclasts. Accordingly, opposite effects were observed in the absence of functional Rho (Chellaiah et al., 2000).

Other small GTPase family members (Ras, Rab) also appear to have important roles in physiological osteoclast function (for a detailed review, see Coxon and Rogers, 2003). Ras GTPases may be essential for osteoclast survival as dominant negative forms of Ras-induced apoptosis in murine osteoclasts (Miyazaki et al., 2000). Moreover, there is evidence to suggest that the downstream mitogen activated protein kinase (MAPK) cascades (MEK/ERK) are required for normal osteoclast differentiation (Miyazaki et al., 2000; Lee et al., 2002), although reports to the contrary have also been published (Matsumoto et al., 2000; Hotokezaka et al., 2002). Rab proteins are likely to be essential for normal osteoclast function, given their role in vesicular transport – a mechanism employed by the ruffled border in the acidification of resorption lacunae. Rab 3b/c and Rab 7 are both expressed in osteoclasts, with the latter being localised specifically to the ruffled border. Furthermore, inhibition of Rab 7 by antisense was shown by Zhao and co-workers to inhibit bone resorption *in vitro* (Zhao et al., 2001). Taken together, these data extend the list of small GTPases that have specific effects on osteoclast function. Delineating how each of these signalling pathways interact under physiological conditions to regulate bone resorption provides the basis for exciting future experiments.

CONCLUDING REMARKS

Specific signalling pathways that regulate multiple aspects of cell differentiation and cell physiology are being uncovered very rapidly by many laboratories. The pronounced effects of several bacterial toxins on the differentiation and activity of the two major bone cell populations, osteoblasts and osteoclasts, have provided researchers with valuable insights for the elucidation of signalling pathways utilised by small GTPases during the process of bone remodelling. This has highlighted the usefulness of bacterial protein toxins as tools to perturb specific signalling pathways in bone cells, resulting in altered cell behaviour. Elucidating further the mechanism of action and specificity of such bacterial toxins in osteoblasts and osteoclasts will undoubtedly provide critical insights for understanding their roles in diseases of the skeleton, and ultimately, for the development of novel therapeutic strategies.

REFERENCES

Ackermann M R, Adams D A, Gerken L L, Beckman M J, and Rimler R B (1993). Purified *Pasteurella multocida* protein toxin reduces acid phosphatase-positive osteoclasts in the ventral nasal concha of gnotobiotic pigs. *Calcif. Tissue Int.*, **52**, 455–459.

Ackermann M R, Rimler R B, and Thurston J R (1991). Experimental model of atrophic rhinitis in gnotobiotic pigs. *Infect. Immun.*, **59**, 3626–3629.

Anderson D M, Maraskovsky E, Billingsley W L, Dougall W C, Tometsko M E, Roux E R, Teepe M C, DuBose R F, Cosman D, and Galibert L (1997). A homologue of the TNF receptor and its ligand enhance T-cell growth and dendritic-cell function. *Nature*, **390**, 175–179.

Athanasou N A (1996). Cellular biology of bone-resorbing cells. *J. Bone. Joint. Surg. Am.*, **78**, 1096–1112.

Aubin J E (1998). Advances in the osteoblast lineage. *Biochem. Cell Biol.*, **76**, 899–910.

Bishop A L and Hall A (2000). Rho GTPases and their effector proteins. *Biochem. J.*, **348**, 241–255.

Blair H C, Teitelbaum S L, Ghiselli R, and Gluck S (1989). Osteoclastic bone resorption by a polarized vacuolar proton pump. *Science*, **245**, 855–857.

Boquet P (2000). The cytotoxic necrotizing factor 1 (CNF1). From uropathogenic *Escherichia Coli. Adv. Exp. Med. Biol.*, **485**, 45–51.

Boyce B F, Yoneda T, Lowe C, Soriano P, and Mundy G R (1992). Requirement of pp 60c-src expression for osteoclasts to form ruffled borders and resorb bone in mice. *J. Clin. Invest.*, **90**, 1622–1627.

Chellaiah M A, Soga N, Swanson S, McAllister S, Alvarez U, Wang D, Dowdy S F, and Hruska K A (2000). Rho-A is critical for osteoclast podosome organization, motility, and bone resorption. *J. Biol. Chem.*, **275**, 11993–12002.

Coxon F P and Rogers M J (2003). The Role of Prenylated Small GTP-Binding Proteins in the Regulation of Osteoclast Function. *Calcif. Tissue Int.*, **72**, 80–84.

Darnay B G, Haridas V, Ni J, Moore P A, and Aggarwal B B (1998). Characterization of the intracellular domain of receptor activator of NF-κB (RANK). Interaction with tumor necrosis factor receptor- associated factors and activation of NF-κB and c-Jun N-terminal kinase. *J. Biol. Chem.*, **273**, 20551–20555.

Darnay B G, Ni J, Moore P A, and Aggarwal B B (1999). Activation of NF-κB by RANK requires tumor necrosis factor receptor-associated factor (TRAF) 6 and NF-κB-inducing kinase. Identification of a novel TRAF6 interaction motif. *J Biol. Chem.*, **274**, 7724–7731.

Ducy P, Schinke T, and Karsenty G (2000). The osteoblast: a sophisticated fibroblast under central surveillance. *Science*, **289**, 1501–1504.

Duong L T, Lakkakorpi P, Nakamura I, and Rodan G A (2000). Integrins and signaling in osteoclast function. *Matrix Biol.*, **19**, 97–105.

Etienne-Manneville S and Hall A (2002). Rho GTPases in cell biology. *Nature*, **420**, 629–635.

Felix R, Fleisch H, and Frandsen P L (1992). Effect of *Pasteurella multocida* toxin on bone resorption in vitro. *Infect. Immun.*, **60**, 4984–4988.

Flatau G, Lemichez E, Gauthier M, Chardin P, Paris S, Fiorentini C, and Boquet P (1997). Toxin-induced activation of the G protein p21 Rho by deamidation of glutamine. *Nature*, **387**, 729–733.

Fujikawa Y, Sabokbar A, Neale S D, Itonaga I, Torisu T, and Athanasou N A (2001). The effect of macrophage-colony stimulating factor and other humoral factors (interleukin-1, -3, -6, and -11, tumor necrosis factor-alpha, and granulocyte macrophage-colony stimulating factor) on human osteoclast formation from circulating cells. *Bone*, **28**, 261–267.

Grigoriadis A E, Wang Z Q, Cecchini M G, Hofstetter W, Felix R, Fleisch H A, and Wagner E F (1994). c-Fos: A key regulator of osteoclast-macrophage lineage determination and bone remodeling. *Science*, **266**, 443–448.

Gwaltney S M, Galvin R J, Register K B, Rimler R B, and Ackermann M R (1997). Effects of *Pasteurella multocida* toxin on porcine bone marrow cell differentiation into osteoclasts and osteoblasts. *Vet. Pathol.*, **34**, 421–430.

Harmey D, Stenbeck G, Nobes C D, Lax A J, and Grigoriadis A E (2004). Regulation of osteoblast differentiation by *Pasteurella multocida* toxin (PMT): A role for Rho GTPase in bone formation. *J. Bone Miner. Res.*, **19**, 661–670.

Helfrich M H and Horton M A (1999). Integrins and adhesion molecules. In *Dynamics of Bone and Cartilage Metabolism*, ed. M J Siebel, S P Robins, and J P Bileziken, pp. 111–125, Academic Press, San Diego.

Henderson B and Nair S P (2003). Hard labour: Bacterial infection of the skeleton. *Trends Microbiol.*, **11**, 570–577.

Hofbauer L C and Heufelder A E (1998). Osteoprotegerin and its cognate ligand: A new paradigm of osteoclastogenesis. *Eur. J. Endocrinol.*, **139**, 152–154.

Horiguchi Y, Inoue N, Masuda M, Kashimoto T, Katahira J, Sugimoto N, and Matsuda M (1997). *Bordetella bronchiseptica* dermonecrotizing toxin induces reorganization of actin stress fibers through deamidation of Gln-63 of the GTP-binding protein Rho. *Proc. Natl. Acad. Sci. USA*, **94**, 11623–11626.

Horiguchi Y, Okada T, Sugimoto N, Morikawa Y, Katahira J, and Matsuda M (1995). Effects of *Bordetella bronchiseptica* dermonecrotizing toxin on bone formation in calvaria of neonatal rats. *FEMS Immunol. Med. Mic.*, **12**, 29–32.

Hotokezaka H, Sakai E, Kanaoka K, Saito K, Matsuo K, Kitaura H, Yoshida N, and Nakayama K (2002). U0126 and PD98059, specific inhibitors of MEK, accelerate differentiation of RAW264.7 cells into osteoclast-like cells. *J. Biol. Chem.*, **277**, 47366–47372.

Hsu H, Lacey D L, Dunstan C R, Solovyev I, Colombero A, Timms E, Tan H L, Elliott G, Kelley M J, Sarosi I, Wang L, Xia X Z, Elliott R, Chiu L, Black T, Scully S, Capparelli C, Morony S, Shimamoto G, Bass M B, and Boyle W J (1999). Tumor necrosis factor receptor family member RANK mediates osteoclast differentiation and activation induced by osteoprotegerin ligand. *Proc. Natl. Acad. Sci. USA*, **96**, 3540–3545.

Iotsova V, Caamano J, Loy J, Yang Y, Lewin A, and Bravo R (1997). Osteopetrosis in mice lacking NF-κB1 and NF-κB2. *Nat. Med.*, **3**,1285–1289.

Jutras I, and Martineau-Doize B (1996). Stimulation of osteoclast-like cell formation by *Pasteurella multocida* toxin from hemopoietic progenitor cells in mouse bone marrow cultures. *Can. J. Vet. Res.*, **60**, 34–39.

Karsenty G, and Wagner E F (2002). Reaching a genetic and molecular understanding of skeletal development. *Dev. Cell*, **2**, 389–406.

Kimman T G, Lowik C W, van de Wee-Pals L J, Thesingh C W, Defize P, Kamp E M, and Bijvoet O L (1987). Stimulation of bone resorption by inflamed nasal mucosa, dermonecrotic toxin-containing conditioned medium from *Pasteurella multocida*, and purified dermonecrotic toxin from *P. multocida*. *Infect. Immun.*, **55**, 2110–2116.

Lacerda H M, Pullinger G D, Lax A J, and Rozengurt E (1997). Cytotoxic necrotizing factor 1 from *Escherichia coli* and dermonecrotic toxin from *Bordetella bronchiseptica* induce p21(rho)-dependent tyrosine phosphorylation of focal adhesion kinase and paxillin in Swiss 3T3 cells. *J. Biol. Chem.*, **272**, 9587–9596.

Lacey D L, Timms E, Tan H L, Kelley M J, Dunstan C R, Burgess T, Elliott R, Colombero A, Elliott G, Scully S, Hsu H, Sullivan J, Hawkins N, Davy E, Capparelli C, Eli A, Qian Y X, Kaufman S, Sarosi I, Shalhoub V, Senaldi G, Guo J, Delaney J, and Boyle W J (1998). Osteoprotegerin ligand is a cytokine that regulates osteoclast differentiation and activation. *Cell*, **93**,165–176.

Lax A J, and Chanter N (1990). Cloning of the toxin gene from *Pasteurella multocida* and its role in atrophic rhinitis. *J. Gen. Microbiol.*, **136**, 81–87.

Lee S E, Woo K M, Kim S Y, Kim H M, Kwack K, Lee Z H, and Kim H H (2002). The phosphatidylinositol 3-kinase, p38, and extracellular signal-regulated kinase pathways are involved in osteoclast differentiation. *Bone*, **30**, 71–77.

Lerm M, Schmidt G, and Aktories K (2000). Bacterial protein in toxins targetting rho GTPases. *FEMS Microbiol. Lett.*, **188**, 1–6.

Lerm M, Schmidt G, Goehring U M, Schirmer J, and Aktories K (1999). Identification of the region of rho involved in substrate recognition by *Escherichia coli* cytotoxic necrotizing factor 1 (CNF1). *J. Biol. Chem.*, **274**, 28999–29004.

Lomaga M A, Yeh W C, Sarosi I, Duncan G S, Furlonger C, Ho A, Morony S, Capparelli C, Van G, Kaufman S, van der Heiden A, Itie A, Wakeham A, Khoo W, Sasaki T, Cao Z D, Penninger J M, Paige C J, Lacey D L, Dunstan C R, Boyle W J, Goeddel D V, and Mak T W (1999). TRAF6 deficiency results in osteopetrosis and defective interleukin-1, CD40, and LPS signaling. *Gene Dev.*, **13**, 1015–1024.

Lowe C, Yoneda T, Boyce B F, Chen H, Mundy G R, and Soriano P (1993). Osteopetrosis in Src-deficient mice is due to an autonomous defect of osteoclasts. *Proc. Natl. Acad. Sci. USA*, **90**, 4485–4489.

Mackay D J and Hall A (1998). Rho GTPases. *J. Biol. Chem.*, **273**, 20685–20688.

Manolagas S C (2000). Birth and death of bone cells: Basic regulatory mechanisms and implications for the pathogenesis and treatment of osteoporosis. *Endocr. Rev.*, **21**, 115–137.

Martineau-Doize B, Caya I, Gagne S, Jutras I, and Dumas G (1993). Effects of *Pasteurella multocida* toxin on the osteoclast population of the rat. *J. Comp. Pathol.*, **108**, 81–91.

Matsumoto M, Sudo T, Saito T, Osada H, and Tsujimoto M (2000). Involvement of p38 mitogen-activated protein kinase signaling pathway in osteoclastogenesis mediated by receptor activator of NF-κB ligand (RANKL). *J. Biol. Chem.*, **275**, 31155–31161.

Miyazaki T, Katagiri H, Kanegae Y, Takayanagi H, Sawada Y, Yamamoto A, Pando M P, Asano T, Verma I M, Oda H, Nakamura K, and Tanaka S (2000). Reciprocal role of ERK and NF-κB pathways in survival and activation of osteoclasts. *J. Cell Biol.*, **148**, 333–342.

Mullan P B and Lax A J (1996). *Pasteurella multocida* toxin is a mitogen for bone cells in primary culture. *Infect. Immun.*, **64**, 959–965.

Mullan P B and Lax A J (1998). *Pasteurella multocida* toxin stimulates bone resorption by osteoclasts via interaction with osteoblasts. *Calcified. Tissue Int.*, **63**, 340–345.

Mundy G R (1998). Bone remodelling. In *Primer on the metabolic bone diseases and disorders of mineral metabolism*, 4th ed., ed. M J Favus, pp. 30–39, Lippincott Williams & Wilkins, Philadelphia.

Nair S P, Meghji S, Wilson M, Reddi K, White P, and Henderson B (1996). Bacterially induced bone destruction: Mechanisms and misconceptions. *Infect. Immun.*, **64**, 2371–2380.

Nesbitt S A and Horton M A (1997). Trafficking of matrix collagens through bone-resorbing osteoclasts. *Science*, **276**, 266–269.

Nobes C D and Hall A (1995). Rho, rac, and cdc42 GTPases regulate the assembly of multimolecular focal complexes associated with actin stress fibers, lamellipodia, and filopodia. *Cell*, **81**, 53–62.

Raisz L G (1999). Physiology and pathophysiology of bone remodeling. *Clin. Chem.*, **45**, 1353–1358.

Rodan G A and Martin T J (1981). Role of osteoblasts in hormonal control of bone resorption–A hypothesis. *Calcif. Tissue Int.*, **33**, 349–351.

Rozengurt E, Higgins T, Chanter N, Lax A J, and Staddon J M (1990). *Pasteurella multocida* toxin: Potent mitogen for cultured fibroblasts. *Proc. Natl. Acad. Sci. USA*, **87**, 123–127.

Salo J, Lehenkari P, Mulari M, Metsikko K, and Vaananen H K (1997). Removal of osteoclast bone resorption products by transcytosis. *Science*, **276**, 270–273.

Sarma U and Flanagan A M (1996). Macrophage colony-stimulating factor induces substantial osteoclast generation and bone resorption in human bone marrow cultures. *Blood*, **88**, 2531–2540.

Schmidt G, Goehring U M, Schirmer J, Lerm M, and Aktories K (1999). Identification of the C-terminal part of *Bordetella dermonecrotic* toxin as a transglutaminase for rho GTPases. *J. Biol. Chem.*, **274**, 31875–31881.

Schmidt G, Sehr P, Wilm M, Selzer J, Mann M, and Aktories K (1997). Gln 63 of Rho is deamidated by *Escherichia coli* cytotoxic necrotizing factor-1. *Nature*, **387**, 725–729.

Schmidt G, Selzer J, Lerm M, and Aktories K (1998). The Rho-deamidating cytotoxic necrotizing factor 1 from *Escherichia coli* possesses transglutaminase activity. Cysteine 866 and histidine 881 are essential for enzyme activity. *J. Biol. Chem.*, **273**, 13669–13674.

Simonet W S, Lacey D L, Dunstan C R, Kelley M, Chang M S, Luthy R, Nguyen H Q, Wooden S, Bennett L, Boone T, Shimamoto G, DeRose M, Elliott R, Colombero A, Tan H L, Trail G, Sullivan J, Davy E, Bucay N, Renshaw-Gegg L, Hughes T M, Hill D, Pattison W, Campbell P, and Boyle W J (1997). Osteoprotegerin: A novel secreted protein involved in the regulation of bone density. *Cell*, **89**, 309–319.

Soriano P, Montgomery C, Geske R, and Bradley A (1991). Targeted disruption of the c-src proto-oncogene leads to osteopetrosis in mice. *Cell*, **64**, 693–702.

Sterner-Kock A, Lanske B, Uberschar S, and Atkinson M J (1995). Effects of the *Pasteurella multocida* toxin on osteoblastic cells in vitro. *Vet. Pathol.*, **32**, 274–279.

Suda T, Nakamura I, Jimi E, and Takahashi N (1997). Regulation of osteoclast function. *J. Bone Miner. Res.*, **12**, 869–879.

Suda T, Takahashi N, and Martin T J (1992). Modulation of osteoclast differentiation. *Endocr. Rev.*, **13**, 66–80.

Suda T, Takahashi N, Udagawa N, Jimi E, Gillespie M T, and Martin T J (1999). Modulation of osteoclast differentiation and function by the new members of the tumor necrosis factor receptor and ligand families. *Endocr. Rev.*, **20**, 345–357.

Takahashi N, Akatsu T, Udagawa N, Sasaki T, Yamaguchi A, Moseley J M, Martin T J, and Suda T (1988). Osteoblastic cells are involved in osteoclast formation. *Endocrinology*, **123**, 2600–2602.

Takai Y, Sasaki T, and Matozaki T (2001). Small GTP-binding proteins. *Physiol. Rev.*, **81**, 153–208.

Teitelbaum S L (2000). Bone resorption by osteoclasts. *Science*, **289**, 1504–1508.

Vaananen H K and Horton M (1995). The osteoclast clear zone is a specialized cell-extracellular matrix adhesion structure. *J. Cell Sci.*, **108**, 2729–2732.

Ward P N, Miles A J, Sumner I G, Thomas L H, and Lax A J (1998). Activity of the mitogenic *Pasteurella multocida* toxin requires an essential C-terminal residue. *Infect. Immun.*, **66**, 5636–5642.

Wei S, Teitelbaum S L, Wang M W, and Ross F P (2001). Receptor activator of nuclear factor-kappa b ligand activates nuclear factor-κB in osteoclast precursors. *Endocrinology*, **142**, 1290–1295.

Yasuda H, Shima N, Nakagawa N, Yamaguchi K, Kinosaki M, Mochizuki S, Tomoyasu A, Yano K, Goto M, Murakami A, Tsuda E, Morinaga T, Higashio K, Udagawa N, Takahashi N and Suda T (1998). Osteoclast differentiation factor is a ligand for osteoprotegerin/osteoclastogenesis-inhibitory factor and is identical to TRANCE/RANKL. *Proc. Natl. Acad. Sci. USA*, **95**, 3597–3602.

Zhang D, Udagawa N, Nakamura I, Murakami H, Saito S, Yamasaki K, Shibasaki Y, Morii N, Narumiya S, and Takahashi N (1995). The small GTP-binding protein, rho p21, is involved in bone resorption by regulating cytoskeletal organization in osteoclasts. *J. Cell Sci.*, **108**, 2285–2292.

Zhao H, Laitala-Leinonen T, Parikka V, and Vaananen H K (2001). Downregulation of small GTPase Rab7 impairs osteoclast polarization and bone resorption. *J. Biol. Chem.*, **276**, 39295–39302.

Zywietz A, Gohla A, Schmelz M, Schultz G, and Offermanns S (2001). Pleiotropic effects of *Pasteurella multocida* toxin are mediated by Gq-dependent and -independent mechanisms: Involvement of Gq but not G11. *J. Biol. Chem.*, **276**, 3840–3845.

CHAPTER 8

Helicobacter pylori mechanisms for inducing epithelial cell proliferation

Michael Naumann and Jean E Crabtree

Helicobacter pylori, the first bacterium to be designated a Class I carcinogen, has a major aetiological role in human gastric carcinogenesis. *H. pylori* infection is acquired primarily in childhood and, in the majority of instances, infection and associated chronic gastritis are lifelong. A key feature of *H. pylori* infection of relevance to the associated increased risk of developing gastric cancer is the hyperproliferation of gastric epithelial cells induced by the bacterium. Infection is associated with increased gastric epithelial cell proliferation in both humans and in experimental animal models.

Clinically, there is a marked diversity in the outcome of *H. pylori* infection and only a few infected subjects will develop gastric cancer (reviewed Peek and Blaser, 2002). Recent studies in Japan show that the risk of cancer with *H. pylori* infection is greatest in infected subjects with nonulcer dyspesia or gastric ulceration who develop severe gastric atrophy and intestinal metaplasia (Uemura et al., 2001). Bacterial virulence factors such as the *cag* pathogenicity island (PAI) (Blaser et al., 1995; Kuipers et al., 1995; Webb et al., 1999) and genetic polymorphisms in the interleukin-1β and IL-1 receptor antagonist genes associated with overexpression of IL-1 and hypochlorhydria (El-Omar et al., 2000; Machado et al., 2001; Furuta et al., 2002) have each been linked to an increased risk of developing gastric atrophy and/or intestinal type gastric cancer.

H. pylori is one of several chronic infections that have recently been associated with the development of neoplasia (see Chapter 9). The cellular and molecular pathways by which *H. pylori* infection promotes epithelial hyperproliferative responses and transformation are the subject of active investigation. *H. pylori* represents an excellent model system to investigate pathogen-induced epithelial cell signalling pathways of relevance to neoplasia. Recent studies have focused on identifying bacterial and host factors involved in the epithelial hyperproliferative responses. Both *in vitro* bacterial–epithelial co-culture systems, and *in vivo* studies in humans and animal models, have been used to examine bacterial factors involved in the hyperproliferative response and the mechanisms by which *H. pylori* alters cell cycle control in the gastric epithelium.

CLINICAL STUDIES

Many clinical studies have demonstrated that infection with *H. pylori* is associated with increased epithelial cell proliferation in humans (Lynch et al., 1995a; Lynch et al., 1995b; Cahill et al., 1996; Jones et al., 1997; Moss et al., 2001). Increased epithelial proliferation is observed early in the natural history of infection, being present in *H. pylori*–infected children (Jones et al., 1997). Lifelong increased cell turnover is likely to be an important risk factor for the development of gastric cancer. Patients with *H. pylori* negative gastritis do not have increased gastric epithelial cell proliferation compared to uninfected controls (Lynch et al., 1995a; Lynch et al., 1995b; Cahill et al., 1996; Panella et al., 1996). Epithelial proliferation indices have been positively correlated with the degree of histological inflammation in both the antrum and corpus mucosa in *H. pylori*–infected patients (Peek et al., 1997; Lynch et al., 1999; Moss et al., 2001). However, stepwise multiple regression analysis has indicated that the only independent predictor of epithelial cell proliferation is the density of *H. pylori* colonisation (Lynch et al., 1999). This suggests that in humans the enhanced epithelial proliferation observed with infection is promoted both as a consequence of the inflammatory response, and by a route independent of inflammation.

The effect of eradication of *H. pylori* infection on gastric epithelial cell proliferation has also been investigated. A significant decrease in gastric epithelial cell proliferation has been observed following successful eradication of *H. pylori* (Brenes et al., 1993; Fraser et al., 1994; Lynch et al., 1995b; Cahill et al., 1995; Leung et al., 2000). In addition, attenuated levels of epithelial cell proliferation have also been observed in patients where *H. pylori* eradication was unsuccessful (Fraser et al., 1994; Lynch et al., 1995b). This may be due to a decrease in the intensity of inflammation and/or bacterial density after therapy in those patients in whom treatment failed. Only one study to date has failed to demonstrate a reduction in gastric epithelial cell proliferation after eradication of *H. pylori* infection (El-Zimaity et al., 2000). Divergent results may be attributed to varying patient populations, labelling techniques, and/or treatment with pharmacological agents such as proton pump inhibitors.

MICHAEL NAUMANN AND JEAN E CRABTREE

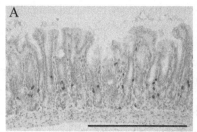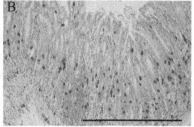

Figure 8.1. Gastric epithelial cell proliferation in the antrum of Mongolian gerbils infected with *H. pylori* SS1 strain. Epithelial cell proliferation in the antral gastric mucosa of A) uninfected control Mongolian gerbil, B) Mongolian gerbil, 4 weeks post-infection with *H. pylori*. Bar = 500 μm. Mongolian gerbils were injected 1 hour prior to sacrifice with bromodeoxyuridine (BrdU), and the presence of proliferating epithelial cells was detected by immunohistochemistry using a monoclonal antibody to BrdU. Adapted from Court et al., 2002.

ANIMAL MODELS

Experimental infection with *H. pylori*, or the related gastric *Helicobacter* sp., *H. mustelae*, or *H. felis*, increases gastric epithelial cell proliferation in a variety of animal species, including Japanese monkeys (Fujiyama et al., 1995), ferrets (Yu et al., 1995), Mongolian gerbils (Peek et al., 2000; Israel et al., 2001; Court et al., 2002) (Figure 8.1), and mice (Fox et al., 1996; Wang et al., 1998; Fox et al., 1999). Dietary factors such as salt intake have also been shown to increase gastric epithelial cell proliferation in the *H. pylori* murine model (Fox et al., 1999).

The most detailed studies to date on *Helicobacter*-induced gastric epithelial cell proliferation have been undertaken in the murine *H. felis* model. *H. felis* infection in mice causes severe inflammation and marked epithelial hyperplasia (Wang et al., 1998; Ferrero et al., 2000). In C57BL/6 mice, *H. felis* induces greater gastric epithelial cell proliferative responses than *H. pylori* (Court et al., 2002). Both the severity of the gastritis induced by *H. felis* infection (Sakagami et al., 1996), and the extent of epithelial proliferation and apoptosis (Wang et al., 1998), are dependent on the strain of mouse infected. In C57BL/6 mice, *H. felis* induces extensive epithelial hyperproliferation in the corpus mucosa, with parietal and chief cells being replaced with mucous secreting cells (Sakagami et al., 1996). Wang et al. (1998) demonstrated that both epithelial proliferative responses and apoptosis regulation are increased in *H. felis* infected C57BL/6 mice, which they considered could relate to the absence of phospholipase A2 in these mice (Wang et al., 1998). Interestingly, detailed analysis of *H. felis* stimulated gastric pathology, epithelial cell proliferation, and apoptosis regulation over a one-year period in C57BL/6 mice has revealed marked gender differences in responses (Court et al., 2003). Significant increases in epithelial proliferation and apoptosis in response to *H. felis* were observed only in female mice, possibly reflecting sex differences in the immune responses and cytokine production. Gender differences in ethanol-induced ulceration and gastritis have been previously observed in rats, with females having increased levels of gastric epithelial cell proliferation compared to males (Liu et al., 2001). The functional importance of gender should be considered in future murine studies on *H. felis–* and *H. pylori–*induced chronic gastritis.

    H. pylori infection is associated with elevated plasma gastrin (El-Omar et al., 1997), a protein known to promote gastric epithelial hyperproliferation (Wang et al., 1996; Tsutsui et al., 1997; Miyazaki et al., 1999). In gerbils, *H. pylori*–induced epithelial cell proliferation has been correlated with elevated serum gastrin levels (Peek et al., 2000). In a transgenic mouse model that over-expresses gastrin, *H. felis* infection accelerates the development of gastric adenocarcinoma (Wang et al., 2000b). However, the effects of *H. felis* infection on gastric epithelial cell proliferation in wild-type and hypergastrinaemic mice have not been examined. The contributions of gastric *Helicobacter* infection, hypergastrinaemia, and the chronic inflammatory response to epithelial hyperproliferation are areas of active research.

H. PYLORI VIRULENCE FACTORS

H. pylori is a genomically diverse pathogen (Suerbaum et al., 1998), and several bacterial virulence factors are now considered to have a key role on the epithelial response to infection. It has become apparent in recent years that the epithelial cellular response to *H. pylori* is variable. Only strains containing the 40 kb *cag* PAI (Censini et al., 1996; Akopypants et al., 1998) trigger signalling cascades in gastric epithelial cells resulting in AP-1 and NF-κB activation (Naumann et al., 1999; Foryst-Ludwig and Naumann, 2000) and multiple associated changes in gene expression. Of particular interest has been the observation that chemokines such as IL-8 are upregulated in gastric epithelial cells by *cag* PAI positive *H. pylori* strains (Crabtree et al., 1994a; Sharma et al., 1995). The upregulation of IL-8 and other neutrophil chemotactic C-X-C chemokines, such as GRO-α (Crabtree et al., 1994b; Eck et al. 2000), in human gastric epithelial cells is probably critical to the association among infection with *cag* PAI positive strains, neutrophilic responses, and more severe gastroduodenal disease (Crabtree et al., 1991; Weel et al., 1996).

The gene products of the *cag* PAI encode a type IV secretory system that translocates CagA (Segal et al., 1999; Asahi et al., 2000; Backert et al., 2000; Odenbreit et al., 2000; Stein et al., 2000) and presumably unknown factors into the gastric epithelial cells. The profile of gene expression, as determined by cDNA array analysis, in gastric epithelial cells differs markedly after culture with wild-type *cag* PAI positive and wild-type *cag* PAI negative strains, with many genes involved with cell cycle control and apoptosis being differentially expressed (Cox et al., 2001). There has been considerable interest in whether *in vivo* specific *H. pylori* strains are associated with enhanced epithelial proliferation and apoptosis.

Evidence for *H. pylori* Strain Related Differences in Gastric Epithelial Cell Proliferation *in vivo*

The contribution of bacterial factors to the induction of gastric epithelial cell proliferation and the regulation of apoptosis is under active investigation. One approach has been to examine the effects of *H. pylori* strains of specific genotype and isogenic mutants on proliferation and apoptosis in gastric epithelial cell lines *in vitro* (Peek et al., 1999). However, there is marked variability in the expression of cell cycle regulatory proteins such as $p27^{KIP1}$ and cyclin E/cdk2 activity induced by *H. pylori* in different gastric epithelial cell lines (Sommi et al., 2002), making extrapolation of *in vitro* studies to events *in vivo* difficult. In addition, the majority of studies have documented inhibition of proliferation by *H. pylori in vitro* (Shirin et al., 1999; Peek et al., 1999), which will not necessarily reflect events *in vivo*.

It is currently unclear from the clinical studies carried out whether gastric epithelial cell proliferation varies according to the *cag* PAI status of the infecting strain. Two studies have reported that gastric epithelial cell proliferation is greater in patients infected with *cagA* positive strains than *cagA* negative strains (Peek et al., 1997; Rokkas et al., 1999), although another study in patients with non-ulcer dyspepsia failed to confirm these observations (Moss et al., 2001). A recent study in a Chinese population where 98% of patients were infected with *cagA+ H. pylori* strains identified increased epithelial cell proliferation in those infected with strains expressing the blood group antigen binding adhesin *babA2* (Yu et al., 2002). The presence or absence of adhesins may account for earlier discrepancies.

There have also been similarly divergent results with respect to the effects of the *cag* PAI on apoptosis in human gastric epithelial cells. Two studies reported that apoptosis was greater in patients infected with *cagA* negative strains than *cagA+* strains (Peek et al., 1997; Rokkas et al., 1999), whilst one reported the converse (Moss et al., 2001). Lipopolysaccharide (LPS) of *H. pylori* *cag* PAI positive strains induced apoptosis in isolated guinea pig–derived gastric pit cells *in vitro*, whereas LPS from *cag* PAI negative strains did not induce apoptosis (Kawahara et al., 2001). Other recent *in vitro* studies indicate that *H. pylori* induces apoptosis via NF-κB dependant cascades (Gupta et al., 2001) – observations not in accordance with the observed reduced apoptosis in the gastric mucosa of patients with *cag* PAI positive strains compared to those with *cag* PAI negative strains (Peek et al., 1997; Rokkas et al., 1999). Activation of the peroxisome proliferator-activated receptor γ (PPARγ) suppressed NF-κB mediated apoptosis *in vitro* (Gupta et al., 2001), but this inhibition was independent of *cag* PAI status. From the limited clinical data, the effects of bacterial virulence factors, such as the *cag* PAI, on gastric epithelial proliferation and apoptosis *in vivo* are unresolved. There are several confounding factors, including probably variable usage of non-steroidal anti-inflammatory drugs, which will effect apoptosis (Leung et al., 2000).

Infection of animal models with genetically defined *H. pylori* strains is an alternative approach to investigate the importance of the *cag* PAI on gastric epithelial proliferation and apoptosis. Studies in mice have mainly used the SS1 *H. pylori* strain (Lee et al., 1997). Despite being *cagA*+, this strain appears to lack a functional *cag* PAI (Crabtree et al., 2002), and fails to induce IL-8 secretion in human gastric epithelial cells *in vitro* (van Doorn et al., 1999). In the mouse, host-induced changes in *cag* PAI genotype and related functions have also been reported (Sozzi et al., 2001). In addition, recent studies indicate that the mouse is preferentially colonised by *cagA* negative strains which induce reduced inflammatory responses (Philpott et al., 2002), thus making it an unsuitable model for investigating *cag* PAI effects on host responses.

Mongolian gerbils can be successfully colonised with *H. pylori* strains, with or without a functional *cag* pathogenicity island (Wirth et al., 1998; Peek et al., 2000; Israel et al., 2001; Akanuma et al., 2002; Wang et al., 2003) (Figure 8.2). Long-term infection is associated with the development of intestinal metaplasia (Ikeno et al., 1999) and gastric adenocarcinoma (Watanabe et al., 1998). In the gerbil, *cag* PAI strains of *H. pylori* have been shown to induce more severe gastritis (Ogura et al., 2000; Israel et al., 2001; Akanuma et al., 2002) and increased epithelial proliferation and apoptosis regulation (Israel et al., 2001), compared to strains lacking a functional *cag* PAI. Gastric epithelial cell proliferation in the gerbil in response to *H. pylori* SS1 strain is significantly greater than in the mouse (Court et al., 2002). During the early stages of infection in the gerbil, the epithelial hyperproliferative responses are confined to the antrum, but with time, progress to the corpus (Israel et al., 2001; Court et al., 2002), mirroring the pathology and proliferative

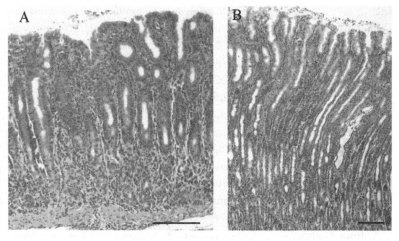

Figure 8.2. Gastric histology in Mongolian gerbil infected with *H. pylori* strain SS1 for 36 weeks. Haematoxylin and Eosin stained sections of A) antral mucosa and B) corpus mucosa. Bar = 100 μm. Adapted from Naumann and Crabtree (2004). (See www.cambridge.org/9780521177467 for color version.)

responses observed with human infection (Moss et al., 2001). The gerbil is thus a useful model to analyse the role of *H. pylori* virulence factors on gastric epithelial proliferation and apoptosis. Further studies with isogenic mutants of other virulence factors will be important to delineate their importance in the epithelial hyperproliferative response.

H. PYLORI INTERFERENCE WITH HOST CELL SIGNALLING

Activation of Proliferation-Associated Kinases ERK/MEK in H. pylori Infection and the Role of COX-2 Induction

While clinical and animal model studies have investigated several aspects of the bacterial-induced hyperproliferative responses, recent *in vitro* studies with gastric epithelial cells have begun to delineate the importance of specific signalling pathways. Furthermore, the contribution of these pathways to over-expression of key genes potentially involved in gastric neoplasia has been examined.

Pathways of great current interest are the induction of nitric oxide synthase and cyclooxygenase 2 (COX-2) in *H. pylori* infection (Fu et al., 1999). As in many human tumour cells, gastric cancer cells over-express COX-2 (Sung et al., 2000), which represents an enzyme responsible for the release of prostaglandin E_2 (PGE_2). PGE_2 is implicated in maintaining the function and structure of the gastric mucosa by modulating diverse cellular functions, such as secretion of fluid and electrolytes, and cell proliferation (Eberhart and Dubois, 1995). One of the mechanisms involved in PGE_2 release is the induction of COX-2 expression. COX-2 mRNA expression and PGE_2 synthesis in MKN28 gastric epithelial cells (Romano et al., 1998) and in human gastric mucosa (Sawaoka et al., 1998; McCarthy et al., 1999; Sung et al., 2000) have been demonstrated in *H. pylori* infection, indicating that COX-2 is involved in *H. pylori*–related gastric pathology. Enhanced transcription of the COX-2 gene in gastric epithelial cells is regulated through a proximal CRE-Ebox enhancer element at −56 to −48 bp by activation of the USF-1 and -2 transcription factors (Jüttner et al., 2003). *H. pylori*–triggered induction of the COX-2 gene appears independent of the *cag* type IV secretion system, and involves activation of the MAP kinases MEK and ERK (Jüttner et al., 2003). A rate-limiting step in the control of PGE_2 is the release of arachidonic acid (AA) from membrane phospholipids, which is known to occur via a number of different pathways. *H. pylori* induces the release of PGE_2 and AA in gastric epithelial cells by activation of the cytosolic phospholipase A_2 (PLA_2) via pertussis toxin-sensitive heterotrimeric $G\alpha_i/G\alpha_o$ proteins and the p38 kinase. PGE_2 production via AA release is predominately synthesised from phosphatidylinositol. In contrast to the *H. pylori* wild-type strain, an isogenic strain with a polar mutation in the *cag* PAI only weakly activates AA synthesis (Pomorski et al., 2001).

An almost complete signalling cascade leading to the activation of the histidine decarboxylase (HDC) promoter in *H. pylori*-infected gastric epithelial cells has recently been described by Wessler et al. (2002). HDC is the key enzyme involved in histamine biosynthesis and converts L-histidine into histamine in enterochromaffin-like cells of the corpus mucosa. Histamine is an important physiological regulator of gastric acid secretion. In contrast to the induction of proinflammatory cytokine-genes such as IL-8 (Li et al., 1999), activation of the HDC promoter in gastric epithelial cells is independent of expression of virulence factors encoded by the *cag* PAI (Wessler et al., 2002). Activation of the HDC promoter involves the activity of the extracellular signal-regulated kinases 1/2 (ERK1/2) and MEK1/2. The hierarchical cascade that leads to MEK/ERK activation in *H. pylori*–infected gastric epithelial cells involves B-Raf, but not c-Raf-1. B-Raf activation in *H. pylori* infection depends on the activity of the Ras-like GTPase Rap1 and the accumulation of cAMP (Wessler et al., 2002). The generation of cAMP in *H. pylori*-infected gastric epithelial cells is also involved in pepsinogen secretion (Jiang et al., 2001). The production of cAMP requires the activity of adenylate cyclase, which is under the control of the heterotrimeric G-protein $G\alpha_s$ (Wessler et al., 2002). It is not currently known how *H. pylori* induces $G\alpha_s$.

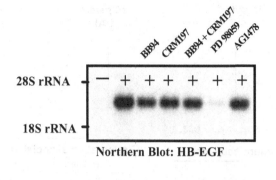

28S rRNA

18S rRNA

Northern Blot: HB-EGF

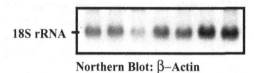

18S rRNA

Northern Blot: β–Actin

Figure 8.3. Northern blot of HB-EGF expression in *H. pylori*–stimulated MKN-1 gastric epithelial cells. *H. pylori*–induced HB-EGF expression is BB94, CRM197, and AG1478 sensitive and completely abolished by PD98059. Northern blot of expression of HB-EGF and *β* actin in MKN-1 gastric epithelial cells 30 minutes following incubation with (+) or without (–) *H. pylori* strain 60190 in the presence or absence BB94, CRM197, AG1478, and PD98059. Adapted from Wallasch et al., 2002.

H. pylori Induces Signalling Cascades via Activation of the Receptor Tyrosine Kinases EGFR and Her2/Neu

The epidermal growth factor receptor (EGFR) and related EGFR ligands are thought to have an important role in gastric mucosal repair (Barnard et al., 1995). Recent studies have demonstrated that *H. pylori* transactivates the EGFR in gastric epithelial cells (Keates et al., 2001; Wallasch et al., 2002). The EGFR transactivation is dependent on extracellular transmembrane metalloprotease cleavage of pro-heparin binding epidermal growth factor (proHB-EGF) and signalling by mature HB-EGF (Wallasch et al., 2002). The upregulation of HB-EGF gene transcription by *H. pylori* requires metalloprotease, EGFR, and Mek1 activities (Wallasch et al., 2002) (Figure 8.3), indicating the involvement of the "triple membrane passing signal" (TMPS) for EGFR transactivation. Previous studies have demonstrated that EGFR transactivation induced by G-protein–coupled receptors (GPCRs) involves triple membrane passing signal (TMPS) transmission events: ligand activation of GPCR, and induction of extracellular transmembrane metalloprotease cleavage of proHB-EGF and EGFR signalling by released mature HB-EGF (Prenzel et al.,

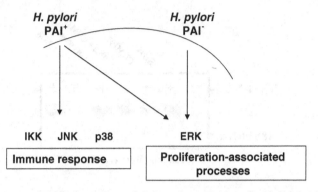

Figure 8.4. *H. pylori* PAI-dependent and -independent activation of epithelial signalling cascades. *H. pylori*–induced signalling by PAI positive strains leads to activation of the IKK, JNK, and p38 kinases, early response transcription factors, and immune response relevant genes. Proliferation-associated activity of the ERK kinases becomes activated by PAI⁺ and PAI⁻ *H. pylori* strains.

1999). EGFR transactivation and increased expression of HB-EGF in gastric epithelial cells are induced by both *cag* PAI positive and *cag* PAI negative *H. pylori* strains (Wallasch et al., 2002) (Figure 8.4). Upregulation of HB-EGF in gastric epithelial cells can be induced by either *H. pylori* cell suspensions or broth culture filtrates, and is independent of a functional *cag* PAI (Romano et al., 1998). One report suggested that EGFR transactivation by *H. pylori* in gastric epithelial cells was dependent on a functional *cag* PAI (Keates et al., 2001), although this was not concordant with earlier observations by the same group that both *cag* positive and *cag* negative *H. pylori* strains stimulated signalling via the ERK/MEK pathway (Keates et al., 1999). The bacterial factors involved in EGFR transactivation remain to be identified. However, induction of TMPS signalling in gastric epithelial cells by *H. pylori* could lead to autocrine/paracrine signalling cascades, promoting enhanced gastric epithelial cell proliferation and decreased apoptosis.

The protease(s) involved in the processing of transmembrane proHB-EGF remains to be identified. Matrix metalloproteinase-3 (MMP-3) cleaves HB-EGF at a specific site in the juxtamembrane domain (Suzuki et al., 1997). *H. pylori* increases MMP-3 in AGS gastric epithelial cells (Gooz et al., 2001) and the bacterium itself has also been reported to have MMP-3 activity (Gooz et al., 2001). Members of the ADAM metalloprotease disintegrin family have also been implicated in ectodomain shedding of pro HB-EGF (Izumi et al., 1998). *H. pylori* upregulates several ADAM genes in cultured gastric epithelial cells (Cox et al., 2001; Yoshimura et al., 2002) and expression of ADAM10

MICHAEL NAUMANN AND JEAN E CRABTREE

and ADAM17 is increased in the gastric mucosa of H. pylori–infected patients (Yoshimura et al., 2002).

Over-expression of key elements of the TMPS cascade in patients with gastric cancer or atrophic gastritis suggests that the EGFR autocrine/paracrine signalling pathway induced by H. pylori is of pathophysiological relevance. H. pylori infection in humans is associated with increased gastric mucosal levels of epidermal growth factor (EGF) protein and EGFR transcripts (Wong et al., 2001). Recent in vitro studies indicate that H. pylori induces the receptor tyrosine kinase HER2/Neu (ErbB-2), another member of the EGF receptor family, in gastric epithelial cells (Churin et al., 2003). Gastric expression of EGFR ligands amphiregulin (Cook et al., 1992; Cox et al., 2001) and HB-EGF (Naef et al., 1996; Murayama et al., 2002) is also increased in patients with H. pylori infection and/or gastric cancer. Additionally, expression of several ADAM genes is strongly increased in gastric cancer mucosa (Yoshimura et al., 2002), raising the possibility that their over-expression may promote amplification of TMPS signalling cascades and dysregulated EGFR transactivation.

H. pylori–induced EGFR transactivation in gastric epithelial cells is likely to be further amplified by the hypergastrinaemia associated with infection. Gastrin similarly increases HB-EGF and also cell proliferation in rodent gastric epithelial cells both in vitro (Miyazaki et al., 1999) and in vivo (Tsutsui et al., 1997; Wang et al., 2000b). The hypergastrinaemia associated with H. pylori infection (El-Omar et al., 1997) and TMPS-dependent EGFR transactivation via GPCR activation (Prenzel et al., 1999) would further promote autocrine or paracrine EGFR transactivation in gastric epithelial cells (Varro et al., 2002). Importantly, when H. pylori infection is lost in patients with severe atrophic gastritis and hypochlorhydria, the hypergastrinaemia may have an important role in promoting EGFR transactivation.

Activation of the c-Met Receptor and the Motogenic Response: A Mechanism of H. pylori–Induced Tumour Development and Invasion

H. pylori induces a phenotype that is known as cell scattering in gastric epithelial cells in vitro (Naumann, 2001). Other phenotypes beside the scattering of H. pylori–infected epithelial cells, like the actin reorganisation with ruffle-like structures, and epithelial cell movement, strictly depend on the functional type IV secretion system, but appear independent of H. pylori CagA protein expression. At the molecular level, these prominent changes in cellular morphology involve the activity of the Rho-GTPases Rac1 and Cdc42 (Churin et al., 2001). Interestingly, the cell spreading and membrane ruffles induced in gastric epithelial cells infected with PAI⁺ *H. pylori* strains are also similar to the changes in cellular morphology induced by hepatocyte growth factor (HGF). The receptor for HGF is the c-Met receptor tyrosine kinase, which is involved in invasive growth of tumour cells upon activation (Matsumoto and Nakamura, 1996). *In vitro*, HGF promotes epithelial cell growth and survival as well as epithelial–mesenchymal transition, where it stimulates the disso-ciation and dispersal of colonies of epithelial cells and the acquisition of a fibroblastic morphology (Thiery, 2002).

Recent data show that *H. pylori* induces the activation of c-Met in AGS gastric epithelial cells (Churin et al., 2003). Epithelial cell clusters become migratory after infection with *H. pylori*. Comparison of the same epithelial cell colonies before, and 4 hours after, *H. pylori* infection demonstrated the strong stimulation of AGS cell motility (Churin et al., 2003). One of the biological responses to EGFR activation is stimulation of cell motility (Xie et al., 1998). However, specific inhibitors of EGFR (AG1478), and of the closely related HER2/Neu receptor (AG825), had no effect on the activation of c-Met by *H. pylori*, and in spite of the presence of inhibitors, AGS cells became migratory after infection (Churin et al., 2003). These observations indicated that *H. pylori* induces the activation of c-Met in AGS cells that can lead to the stimulation of host cell motogenic response.

Like HGF, the InlB protein of *Listeria monocytogenes* binds to c-Met, thereby inducing tyrosine phosphorylation of several proteins (Gab1, Cbl and Shc) (Ireton et al., 1998). As demonstrated by Shen et al. (2000), InlB is also able to trigger scattering of some epithelial cell lines. The direct involve-ment of c-Met in the stimulation of host epithelial cell motogenic response by *H. pylori* was confirmed by using small interfering RNA (siRNA) to silence the expression of the c-Met receptor by RNA interference (RNAi) in epithe-lial cells. A siRNA to c-Met efficiently and specifically silenced c-Met receptor expression, without affecting EGF receptor expression. Epithelial cells trans-fected with siRNA to c-Met were resistant to the induction of motility by *H. pylori*. Further, the silencing of c-Met receptor expression had no effect on CagA tyrosine phosphorylation (Churin et al., 2003). Compared to the PAI positive wild-type strain, an isogenic *cagA* mutant strain induced only a weak motogenic response in AGS cells; and a *virB11* mutant strain, which lacks a functional type IV secretion system, also failed to promote the motogenic response. Furthermore, over-expression of CagA in AGS cells did not induce motility, indicating that *H. pylori* infection and translocation of the CagA protein are required for the motogenic response (Churin et al., 2003).

Following activation of c-Met, the multifunctional docking site medi-ates the binding of several adapter proteins that in turn recruit several signal-transducing proteins (Furge et al., 2000). Disruption of the multifunctional docking site abrogates the capability of c-Met to induce oncogenic transformation and invasive growth of tumour cells (Bardelli et al., 1998). Biochemical studies indicate that following translocation CagA targets only the phosphorylated c-Met receptor and that this interaction is independent of CagA phosphorylation (Churin et al., 2003). Physical interaction of CagA and PLCγ and activation of PLCγ by *H. pylori* contribute in the motogenic response. Inhibition of the PLCγ signalling pathway blocks growth factor-induced cell motility (Kassis et al., 2001), and inhibition of PLCγ by using the pharmacological agent U73122 suppresses the motogenic response of AGS cells after *H. pylori* infection.

Based on previous studies, wild-type *H. pylori* strains and the *cagA* mutant strain could activate Rho GTPases Rac1 and Cdc42 in AGS gastric epithelial cells. Furthermore, Rac1 and Cdc42 are recruited to the site of bacterial attachment (Churin et al., 2001). Rho GTPases control polarity, protrusion, and adhesion during cell movement (Nobes and Hall, 1999). Thus, a weak motogenic response of AGS cells to infection with the *cagA* mutant strain could be explained by activation of Rho GTPases that leads to the transient polarisation of the host cells. However, the physical interaction of CagA with PLCγ is necessary to produce the complete motogenic response of AGS cells after *H. pylori* infection.

After binding to the multisubstrate docking site of c-Met, adapter proteins recruit several SH2 domain–containing proteins to form an intricate signalling complex (Furge et al., 2000). One of the proteins that plays an important role in c-Met signalling is the large adapter protein Gab1 (Weidner et al., 1996). Growth factor treatment can induce Gab1 tyrosine phosphorylation and its direct association with the SH2 domains of several signal transducers, including phosphatidylinositol 3-OH kinase (PI3-K), PLCγ, and SHP-2 phosphatase (Weidner et al., 1996). However, CagA does not co-immunoprecipitate with Gab1. Furthermore, another adapter protein Grb2 (Ponzetto et al., 1994) also fails to bind CagA (Higashi et al., 2002b; Churin et al., 2003). The interaction of the tyrosine phosphatase SHP-2 and CagA has been described recently (Higashi et al., 2002a). However, this interaction was demonstrated in AGS cells transfected with the plasmid encoding CagA. Thus, CagA directly interacts with signal-transducing proteins and may play a role as an adapter protein in growth factor receptor signalling.

The dual protein/phospholipid kinase PI3-K has been shown to be activated during growth factor signalling (Comoglio and Boccaccio, 2001; Kassis et al., 2001). *H. pylori* infection activates PI3-K in AGS cells, whereas LY294002 strongly inhibited its activation. However, in spite of the presence of the PI3-K inhibitor, the AGS cells were motile. In contrast to AGS cells,

MDCK cells treated with a specific PI3-K inhibitor and infected with *H. pylori* do not show scattering. As the MDCK cells are polarised primary canine kidney cells, and thus fundamentally different from AGS cells, the observed difference in PI3-K requirement may be due to cell type specificity (Churin et al., 2003).

Studies using MAPK signalling pathway inhibitors have established a role for the MAPK signalling pathway in regulating cell motility (Klemke et al., 1997). Within the family of MAPK the extracellular regulated kinases (ERKs) promote cell motility in a transcription-independent manner (Klemke et al., 1997). It has been previously reported that *H. pylori* activates PAI-independent ERKs in AGS cells (Keates et al., 1999; Wessler et al., 2000). Inhibition of MAP kinases with PD98059 completely blocks ERK activation and the *H. pylori*–induced motogenic response (Churin et al., 2003). These observations demonstrate that MAP kinase signalling events are critical for the induction of the motogenic response in *H. pylori*–infected epithelial cells.

The induction of the motogenic response by *H. pylori* in epithelial cells represents an example of how human microbial pathogens can activate growth factor receptor tyrosine kinases, and modify signal transduction in the cell using translocated bacterial proteins. The *H. pylori* effector protein, CagA, acts inside the cells to target the c-Met receptor and to enhance the motogenic response, which suggests that dysregulation of growth factor receptor signalling could play a role in motility and invasiveness of epithelial cells. The activation of the motogenic response in *H. pylori*–infected epithelial cells suggests that CagA could be involved in tumour progression. Numerous experimental and clinical data indicate a particular role of HGF and the proto-oncogene c-Met in tumour invasive growth. Thus, *H. pylori*–induced c-Met receptor signal transduction pathways could be responsible for cancer onset and tumour progression. Moreover, *H. pylori* colonisation could not only be associated with the development of gastric cancer, but could also promote tumour invasion through stimulation of the motogenic response in infected cells.

HOST CELL FACTORS REGULATING *H. PYLORI*–INDUCED CELL CYCLE CONTROL, APOPTOSIS, AND HYPERPROLIFERATIVE RESPONSES

The contribution of host factors to the induction of gastric epithelial cell proliferation and apoptosis is under active investigation. Experimental infection of *H. felis* in transgenic mice with specific gene deletions has been a useful approach for the identification of potential host contributory factors. Several studies have shown reduced gastric epithelial proliferative responses to infection with gastric *Helicobacter* sp. in immune deficient mice, emphasising the importance of mucosal inflammation in the murine model. $RAG^{-/-}$ mice (Roth et al., 1999), mice with severe combined immune deficiency disorder (SCID) (Symthies et al., 2000) and mice deficient in interferon regulatory factor (Sommer et al., 2001) or gamma interferon (Symthies et al., 2000) do not develop gastritis and associated epithelial hyperproliferative responses. In contrast, $IL\text{-}10^{-/-}$ mice (Berg et al., 1998) and mice lacking functional TGF-β type II receptor (Hahm et al., 2002) develop severe hyperplastic gastritis with infection with gastric *Helicobacter* sp., demonstrating the functional importance of anti-inflammatory and immune regulatory cytokines in bacterially induced epithelial hyperproliferation.

(183)

H. *pylori* infection results in increased expression of Fas antigen on gastric epithelial cells (Rudi et al., 1998), and experimental studies have demonstrated that Fas ligand positive T cells can induce epithelial cell cytotoxicity via Fas/Fas ligand interactions (Wang et al., 2000a). T cells may therefore have an important role in the induction of gastric epithelial cell apoptosis in *H. pylori* infection and thus promote hyperproliferative epithelial responses. In Fas antigen-deficient mice, inflammation induced by *H. felis* is comparable to wild-type mice, but infection does not result in the increases in gastric epithelial cell proliferation or apoptosis observed with wild-type mice (Houghton et al., 2000). In addition, in contrast to wild-type mice, there is no progression to atrophic gastritis in *H. felis* Fas antigen-deficient mice. These studies emphasise the important role of Fas antigen in *Helicobacter* induced gastritis in the murine model, and suggest that in this model gastric epithelial cell proliferation occurs as a consequence of increased Fas–mediated apoptosis.

Transgenic mice have also been used to examine the importance of p53 (Fox et al., 1996) and *Apc* (Fox et al., 1997) on cell cycle changes induced by gastric *H. felis* infection. Infected p53 hemizygous mice displayed higher epithelial cell proliferation than wild-type controls at 12 months, but a later longer term study (Fox et al., 2002) did not confirm increased pathology in p53 hemizygous mice. Interestingly, *H. felis* infection in mice with a truncated *Apc* gene resulted in both decreased inflammation and epithelial cell proliferative responses (Fox et al., 1997) compared to wild-type mice. The reduced proliferative responses to *H. felis* in the mice with a truncated *Apc* gene are likely to relate to impaired inflammatory responses to infection.

A number of studies have analysed alterations in host proteins of relevance to epithelial cell proliferation and apoptosis in *H. pylori*–infected epithelial cells *in vitro*. Studying colonic T84 epithelial cells, Le'Negrate et al. (2001) showed that *H. pylori* triggers apoptosis via a Fas-dependent pathway, which also depends on the expression of the *cag* PAI. Activation of the nuclear hormone transcription factor peroxisome proliferator-activated receptor γ (PPARγ) suppresses *H. pylori*–induced apoptosis, which depends presumably on the ability of PPARγ to inhibit *H. pylori*–induced activation of NF-κB (Gupta et al., 2001).

In human umbilical vein endothelial cells (HUVECs) a *H. pylori* water-soluble extract induced apoptosis and increased phosphorylation of the tumour suppressor p53. Elevated expression of the cell cycle inhibitor p21 and Bax was also observed (Kurosawa et al., 2002). In neutrophils the *H. pylori* water extract inhibited apoptosis and upregulated expression of Bcl-X_L, whereas caspases 8 and 3 were suppressed. The expression of Bax and Bak was upregulated, and Bcl-2, Bcl-X_L, and Mcl-1 downregulated during neutrophilic differentiation (Kim et al., 2001). In contrast, *H. pylori* induced caspases 3, 7, and 8 and anti-apoptotic proteins c-IAP1 and c-IAP2 in epithelial cells in a time-dependent manner, whereas Bax, Bak, and Bcl-X_L were not changed (Maeda et al., 2002).

In vitro studies have demonstrated that Cyclin D1 (Hirata et al., 2001) and c-Fos (Mitsuno et al., 2002) expression induced in *H. pylori*-infected epithelial cells is partly dependent on the *cag* PAI. Cyclin D1 regulates passage through the G_1 phase, and cyclin D1 over-expression shortens the G_1 phase and increases the rate of cellular proliferation. Activation of Cyclin D1 and c-Fos involves mitogen-activated protein kinase/extracellular signal–regulated kinase kinase (MEK) activity (Hirata et al., 2001; Mitsuno et al., 2002).

Viewing the effects of *H. pylori* on growth control, we see that exposure of epithelial cells to *H. pylori* alters cell proliferation rates and apoptosis *in vitro* and *in vivo* (Ebert et al., 2000; Xia and Talley, 2001). *H. pylori* is capable of inhibiting cell cycle progression and induces apoptosis in AGS cells, which is associated with a reduced expression of the cell cycle inhibitor p27^{KIP1} (Shirin et al., 2000). Other reports by Peek et al. (1997, 1999) show that *H. pylori* induces cell cycle progression and apoptosis, which do not affect the expression of p53 or the cell cycle inhibitor p21. Cyclin D3 is frequently detected in the antral mucosa of *H. pylori*–infected patients (Miehlke et al., 2002), and Cyclin D2 overexpression, together with reduced p27^{KIP1} expression, are closely associated with *H. pylori* infection and intestinal metaplasia (Shirin et al., 2000; Yu et al., 2001). Mucosal expression of Cyclin D1, p53, and p21 is significantly higher in patients with intestinal metaplasia (Polat et al., 2002). In addition, expression of the intestine specific homeobox-gene Cdx2 has also been observed in patients with chronic gastritis and is also closely associated with intestinal metaplasia (Satoh et al., 2002). Cdx2 plays an important role in differentiation and maintenance of intestinal epithelial cells.

Presumably, apoptosis in epithelial cells decreases but proliferation increases in the progression to neoplasia in the human gastric mucosa. The role of *H. pylori* virulence factors and the molecular mechanisms contributing to the malignant transformation in the gastric mucosa remain to be clarified.

CONCLUSIONS

Inflammation induced by microbial infection is likely to be a critical component in tumour development. Gastric cancer arises as a consequence of long-term *H. pylori* infection, chronic irritation, and inflammation. Perhaps the best evidence for the significance of inflammation during neoplastic progression comes from the observation of reduced cancer risk among long-term users of aspirin and non-steroidal anti-inflammatory drugs (NSAIDs). The ability of NSAIDs to inhibit cyclo-oxygenases (COX-1 and -2) underlies their mechanism of chemoprevention. Activation of COX-2 expression by *H. pylori* represents just one of several examples how this microorganism contributes to processes of cellular dysregulation. Cancer is a multistep process, which involves alterations of a number of different factors including tumour-suppressor genes, dominant oncogenes, and receptor tyrosine kinases (RTKs). Constitutive activation of EGFR and c-Met in cells of aggressive and invasive tumours has often been described. Downregulation of RTKs may be an applicable approach for the suppression of transforming signalling pathways, and offer novel potential targets for the treatment and/or prevention of malignancies. The recent development of a series of relatively specific receptor tyrosine kinase inhibitors, and their ability to inhibit the proliferation of tumour cells, show that inhibition of deregulated receptor tyrosine kinases is often enough to slow proliferation and tumour progression.

It is essential to deepen our understanding of receptor crosstalk in *H. pylori*–infected epithelium and its contribution to EGFR, Her2/Neu, and c-Met activation. Dysregulated cell motility and metastasis following inappropriate c-Met activation involve collaborations with many other receptors and multiple signalling pathways. In addition, the study of signalling pathways by which EGFR, Her2/Neu, and c-Met expression and activity are regulated in *H. pylori* infection may identify promising therapeutic targets for anticancer drug development. *In vitro* studies have revealed a multitude of mechanisms by which Rho proteins can promote cell proliferation. In *H. pylori*–infected tissue the crucial pathways that link Rho proteins to tumorigenesis remain to be identified, although the stimulation of COX-2 activity by Rho is likely to be involved. To define Rho proteins or effectors, the next step will be to develop viable pharmacological or gene therapy agents to interfere with Rho-protein functions. The use of RNA interference (RNAi) to downregulate the expression of anti-apoptotic genes is a further potential therapeutic approach. Matrix metalloproteases (MMPs) and metalloprotease-disintegrins (ADAMs) expression, which is increased by *H. pylori*, have a pivotal role in modifying the cellular microenvironment and neoplastic tumour progression. It will be important in the future to characterise also the functions of these proteases, and identify the proteases involved in EGFRR transactivation and hyperproliferative epithelial responses.

ACKNOWLEDGEMENTS

Work in the laboratory of JEC is supported by Yorkshire Cancer Research and the European Commission (contract ICA4-CT-1999-10010). The work in the institute of MN is supported by the Deutsche Forschungsgemeinschaft grant Na 292/6-2.

REFERENCES

Akanuma M, Maeda S, Ogura K, Mitsuno Y, Hirata Y, Ikenoue T, Otsuka M, Watanabe T, Yamaji Y, Yoshida H, Kawabe T, Shiratori Y, and Omata M (2002). The evaluation of putative virulence factor of *Helicobacter pylori* for gastroduodenal disease by use of a short term Mongolian gerbil infection model. *J. Infect. Dis.*, **185**, 341–347.

Akopyants N S, Clifton S W, Kersulyte D, Crabtree J E, Youree B E, Reece C A, Bukanov N O, Drazek S E, Roe B A, and Berg D E (1998). Analyses of the *cag* pathogenicity island of *Helicobacter pylori*. *Mol. Microbiol.*, **28**, 37–54.

Asahi M, Azuma T, Ito S, Ito Y, Suto H, Nagai Y, Tsubokawa M, Tohyama Y, Maeda S, Omata M, Suzuki T, and Sasakawa C (2000). *Helicobacter pylori* CagA protein can be tyrosine phosphorylated in gastric epithelial cells. *J. Exp. Med.*, **191**, 593–602.

Backert S, Ziska E, Brinkmann V, Zimny-Arndt U, Fauconnier A, Jungblut P R, Naumann M, and Meyer T F (2000). Translocation of the *Helicobacter pylori* CagA protein in gastric epithelial cells by a type IV secretion apparatus. *Cell. Microbiol.*, **2**, 155–164.

Bardelli A, Longati P, Gramaglia D, Basilico C, Tamagnone L, Giordano S, Ballinari D, Michieli P, and Comoglio P M (1998). Uncoupling signal transducers from oncogenic MET mutants abrogates cell transformation and inhibits invasive growth. *Proc. Natl. Acad. Sci. USA*, **95**, 14379–14383.

Barnard J A, Beauchamp R D, Russell W E, Dubois R N, and Coffey R J (1995). Epidermal growth factor-related peptides and their relevance to gastrointestinal pathophysiology. *Gastroenterology*, **108**, 564–580.

Berg D J, Lynch N A, Lynch R G, and Lauricella D M (1998). Rapid development of severe hyperplastic gastritis with gastric epithelial dedifferentiation in *Helicobacter felis*-infected IL-10–/– mice. *Am. J. Pathol.*, **152**, 1377–1386.

Blaser M J, Perez-Perez G I, Kleanthous H, Cover T L, Peek R M, Chyou P H, Stemmermann G N, and Nomura A (1995). Infection with *Helicobacter pylori* strains possessing cagA is associated with increased risk of developing adenocarcinoma of the stomach. *Cancer Res.*, **55**, 2111–2115.

Brenes F, Ruiz B, Correa P, Hunter F, Rhamakrishnan T, Fontham E, and Shi T Y (1993). *Helicobacter pylori* causes hyperproliferation of the gastric epithelium: Pre- and post-eradication indices of proliferating cell nuclear antigen. *Am. J. Gastroenterol.*, **88**, 1870–1875.

Cahill R J, Xia H, Kilgallen C, Beattie S, Hamilton H, and O'Morain C (1995). Effect of eradication of *Helicobacter pylori* infection on gastric epithelial cell proliferation. *Digest. Dis. Sci.*, **40**, 1627–1631.

Cahill R J, Kilgallen C, Beatti S, Hamilton H, and O'Morain C (1996). Gastric epithelial cell kinetics in the progression from normal mucosa to gastric carcinoma. *Gut*, **38**, 177–181.

Censini S, Lange C, Xiang Z, Crabtree J E, Ghiara P, Borodovsky M, Rappuoli R, and Covacci A (1996). *cag*, a pathogenicity island of *Helicobacter pylori*, encodes Type I-specific and disease-associated virulence factors. *Proc. Natl. Acad. Sci. USA*, **93**, 14648–14653.

Churin Y, Kardalinou E, Meyer T F, and Naumann M (2001). Pathogenicity island-dependent activation of Rho GTPases Rac1 and Cdc42 in *Helicobacter pylori* infection. *Mol. Microbiol.*, **40**, 815–823.

Churin Y, Al-Ghoul L, Kepp O, Meyer T F, Birchmeier W, and Naumann M (2003). *Helicobacter pylori* CagA protein targets the c-Met receptor and enhances the motogenic response. *J. Cell Biol.*, **161**, 249–255.

Comoglio P M and Boccaccio C (2001). Scatter factors and invasive growth. *Semin. Cancer Biol.*, **11**, 153–165.

Cook P W, Pittelkow M R, Keeble W W, Graves-Deal R, Coffey R J, and Shipley G D (1992). Amphiregulin messenger RNA is elevated in psoriatic epidermis and gastrointestinal carcinomas. *Cancer Res.*, **52**, 3224–3227.

Cox J M, Clayton C L, Tomita T, Wallace D M, Robinson P A, and Crabtree J E (2001). cDNA array analysis of *cag* pathogenicity island-associated *Helicobacter pylori* epithelial cell response genes. *Infect. Immun.*, **69**, 6970–6980.

Court M, Robinson P A, Dixon M F, and Crabtree J E (2002). Gastric *Helicobacter* species infection in murine and gerbil models: Comparative analysis of effects of *H. pylori* and *H. felis* on gastric epithelial cell proliferation. *J. Infect. Dis.*, **186**, 1348–1352.

Court M, Robinson P A, Dixon M F, Jeremy A H T, and Crabtree J E (2003). The effect of gender on *Helicobacter felis* mediated gastritis, epithelial cell proliferation and apoptosis in the mouse model. *J. Pathol.*, **201**, 303–311.

Crabtree J E, Taylor J E, Wyatt J I, Heatley R V, Shallcross T M, Tompkins D S, and Rathbone B J (1991). Mucosal IgA recognition of *Helicobacter pylori* 120 kDa protein, peptic ulceration and gastric pathology. *Lancet* **338**, 332–335.

Crabtree J E, Farmery S M, Lindley I J D, Figura N, Peichl P, and Tompkins D S (1994a). CagA/cytotoxic strains of *Helicobacter pylori* and interleukin-8 in gastric epithelial cells. *J. Clin. Pathol.*, **47**, 945–950.

Crabtree J E, Wyatt J I, Trejdosiewicz L K, Peichl P, Nichols P N, Ramsey N, Primrose J N, and Lindley I J D (1994b). Interleukin-8 expression in *Helicobacter pylori*, normal and neoplastic gastroduodenal mucosa. *J. Clin. Pathol.*, **47**, 945–950.

Crabtree J E, Ferrero R L, and Kusters J G (2002). The mouse colonizing *Helicobacter pylori* strain SS1 may lack a functional cag pathogenicity island. *Helicobacter*, **7**, 139–140.

Ebert M P, Yu J, Sung J J, and Malfertheiner P (2000). Molecular alterations in gastric cancer: The role of *Helicobacter pylori*. *Eur. J. Gastroenl. Hepat.*, **12**, 795–798.

Eberhart C E and Dubois R N (1995). Eicosanoids and the gastrointestinal tract. *Gastroenterology*, **109**, 285–301.

Eck M, Schmausser K, Scheller A, Toksoy A, Kraus M, Menzel T, Muller-Hermelink H K, and Gillitzer R (2000). CXC chemokines Groα/IL-8 and IP-10/MIG in *Helicobacter pylori* gastritis. *Clin. Exp. Immunol.*, **122**, 192–199.

El-Omar E, Oien K, El-Nujumi A, Gillen D, Wirz A, Dahill P, Williams C, Ardhill J E, and McColl K E (1997). *H. pylori* infection and chronic gastric acid hyposecretion. *Gastroenterology*, **113**, 15–24.

El-Omar E M, Carrington M, Chow W H, McColl K E, Bream J H, Young H A, Herrera J, Lissowska J, Yuan C C, Rothman N, Lanyon G, Martin M, Fraumeni J F, and Rabkin C S (2000). Interleukin-1 polymorphisms associated with increased risk of gastric cancer. *Nature*, **404**, 398–402.

El-Zimaity H M, Graham D Y, Genta R M, and Lechago J (2000). Sustained increase in gastric antral epithelial cell proliferation despite cure of *Helicobacter pylori* infection. *Am. J. Gastroenterol.*, **95**, 930–935.

Ferrero R L, Ave P, Radcliff F J, Labigne A, and Huerre M R (2000). Outbred mice with long-term *Helicobacter felis* infection develop both gastric lymphoid tissue and glandular hyperplastic lesions. *J. Pathol.*, **191**, 333–340.

Foryst-Ludwig A and Naumann M (2000). p21-activated kinase 1 activates the nuclear factor kappa B (NF-kappa B)-inducing kinase Iκ B kinases NFκ B pathway and proinflammatory cytokines in *Helicobacter pylori* infection. *J. Biol. Chem.*, **275**, 39779–39785.

Fox J G, Li X, Cahill R J, Andrutis K, Rustgi A K, Odze R, and Wang T C (1996). Hypertrophic gastropathy in *Helicobacter felis*-infected wild-type C57BL/6 mice and p53 hemizygous transgenic mice. *Gastroenterology*, **110**, 155–166.

Fox J G, Dangler C A, Whary M T, Edelman W, Kucherlapati R, and Wang T C (1997). Mice carrying a truncated *Apc* gene have diminished gastric epithelial proliferation, gastric inflammation and humoral immunity in response to *Helicobacter felis* infection. *Cancer Res.*, **57**, 3972–3978.

Fox J G, Dangler C A, Taylor N S, King A, Koh T J, and Wang T C (1999). High-salt diet induces gastric epithelial hyperplasia and parietal cell loss, and enhances *Helicobacter pylori* colonization in C57BL/6 mice. *Cancer Res.*, **59**, 4823–4828.

Fox J G, Sheppard B J, Dangler C A, Whary M T, Ihrig M, and Wang T C (2002). Germ-line *p53*-targeted disruption inhibits *Helicobacter*-induced premalignant lesions and invasive gastric carcinoma through down-regulation of Th1 proinflammatory responses. *Cancer Res.*, **62**, 696–702.

Fraser A G, Sim R, Sankey E A, Dhillon A P, and Pounder R E (1994). Effect of eradication of *Helicobacter pylori* on gastric epithelial cell proliferation. *Aliment. Pharm. Therap.*, **8**, 167–173.

Fu S, Ramanujam K S, Wong A, Fantry G T, Drachenberg C B, James S P, Meltzer S J, and Wilson K T (1999). Increased expression and cellular localization of inducible nitric oxide synthase and cyclooxygenase 2 in *Helicobacter pylori* gastritis. *Gastroenterology*, **116**, 1319–1329.

Furge K A, Zhang Y W, and Van de Woude G F (2000). Met receptor tyrosine kinase: Enhanced signaling through adapter proteins. *Oncogene*, **19**, 5582–5589.

Furuta T, El-Omar E M, Xiao F, Shirai N, Takashima M, and Sugimurra H (2002). Interleukin 1β polymorphisms increase risk of hypochlorhydria and atrophic gastritis and reduce risk of duodenal ulcer recurrence in Japan. *Gastroenterology*, **123**, 92–105.

Fujiyama K, Fujioka T, Murakamik K, and Naru M (1995). Effects of *Helicobacter pylori* infection on gastric mucosal defence factors in Japanese monkeys. *J. Gastroenterol.*, **30**, 441–416.

Gooz M, Gooz P, and Smolka A J (2001). Epithelial and bacterial metalloproteinases and their inhibitors in *H. pylori* infection of human gastric epithelial cells. *Am. J. Physiol.*, **281**, G823–G832.

Gupta R A, Polk D B, Krishna U, Israel D A, Yan F, DuBois R N, and Peek R M (2001). Activation of peroxisome proliferator-activated receptor γ suppresses nuclear factor κ B–mediated apoptosis induced by *Helicobacter pylori* in gastric epithelial cells. *J. Biol. Chem.*, **276**, 31059–31066.

Hahm K B, Lee K M M, Kim Y B, Hong W S, Lee W H, Han S U, Kim M W, Ahn B O, Oh T Y, Lee M H, Green J, and Kim S J (2002). Conditional loss of TGF-β signalling leads to increased susceptibility to gastrointestinal carcinogenesis in mice. *Aliment. Pharm. Therap.*, **16** (suppl. 2), 115–127.

Higashi H, Tsutsumi R, Muto S, Sugiyama T, Azuma T, Asaka M and Hatakeyama M (2002a). SHP-2 tyrosine phosphatase as an intracellular target of *Helicobacter pylori* CagA protein. *Science*, **295**, 683–686.

Higashi H, Tsutsumi R, Fujita A, Yamazaki S, Asaka M, Azuma T, and Hatakeyama M (2002b). Biological activity of the *Helicobacter pylori* virulence factor CagA is determined by variation in the tyrosine phosphorylation sites. *Proc. Natl. Acad. Sci. USA*, **99**, 14428–14433.

Hirata Y, Maeda S, Mitsuno Y, Akanuma M, Yamaji Y, Ogura K, Yoshida H, Shiratori Y, and Omata M (2001). *Helicobacter pylori* activates the cyclin D1 gene through mitogen-activated protein kinase pathway in gastric cancer cells. *Infect. Immun.*, **69**, 3965–3971.

Houghton J M, Bloch L M, Goldstein M, von Hagen S, and Korah R M (2000). In vivo disruption of the Fas Pathway abrogates gastric growth alterations secondary to *Helicobacter* infection. *J. Infect. Dis.*, **182**, 856–864.

Ikeno T, Ota H, Sugiyama A, Ishida K, Katsuyama T, Genta R M, and Kwasaki S (1999). *Helicobacter pylori*-induced chronic active gastritis, intestinal metaplasia, and gastric ulcer in Mongolian gerbils. *Am. J. Pathol.*, **154**, 951–960.

Ireton K and Cossart P (1998). Interaction of invasive bacteria with host signaling pathways. *Curr. Opin. Cell Biol.*, **10**, 276–283.

Israel D A, Salama N, Arnold C N, Moss S F, Ando T, Wirth H P, Tham K T, Camorlinga M, Blaser M J, Falkow S, and Peek R M (2001). *Helicobacter pylori* strain specific differences in genetic content, identified by microarray, influence host inflammatory responses. *J. Clin. Invest.*, **107**, 611–620.

Izumi Y, Hirata M, Hasuwa H, Iwamoto R, Umata T, Miyado K, Tamai Y, Kurisaki T, Sehera-Fujisawa A, Ohno S, and Mekada E (1998). A metalloprotease-disintegrin, MDC9/meltrin-γ/ADAM9 and PKCδ are involved in TPA-induced ectodomain shedding of membrane-anchored heparin-binding EGF-like growth factor. *EMBO J.*, **17**, 7260–7272.

Jiang H X, Pu H, Huh N H, Yokota K, Oguma K, and Namba M (2001). *Helicobacter pylori* induces pepsinogen secretion by rat gastric cells in culture via a cAMP signal pathway. *Int. J. Mol. Med.*, **7**, 625–629.

Jones N L, Shannon P T, Cutz E, Yeger H, and Sherman P (1997). Increase in proliferation and apoptosis of gastric epithelial cells early in the natural history of *Helicobacter pylori* infection. *Am. J. Pathol.*, **151**, 1695–1703.

Jüttner S, Cramer T, Wessler S, Walduck A, Schmitz F, Wunder C, Weber M, Fischer S, Wiedenmann B, Meyer T F, Naumann M, and Höcker M (2003). *Helicobacter pylori* infection stimulates cyclooxygenase-2 gene expression in gastric epithelial cells: Essential role of pathogenicity island-independent activation of USF1/-2 and CREB transcription factors. *Cell. Microbiol.*, **5**, 821–834.

Kassis J, Lauffenburger D A, Turner T, and Wells A (2001). Tumor invasion as dysregulated cell motility. *Semin. Cancer Biol.*, **11**, 105–117.

Kawahara T, Kuwano Y, Teshima-Kondo S, Sugiyama T, Kawai T, Nikawa T, Kishi K, and Rokutan K (2001). *Helicobacter pylori* lipopolysaccharide from type I, but not type II strains, stimulates apoptosis of cultured gastric mucosal cells. *J. Med. Invest.*, **48**, 167–174.

Keates S, Keates A C, Warny M, Peek R M, Murray P G, and Kelly C P (1999). Differential activation of mitogen-activated protein kinases in AGS gastric epithelial cells by *cag+* and *cag– Helicobacter pylori*. *J. Immunol.*, **163**, 5552–5559.

Keates S, Sougioultzis S, Keates A, Zhao D, Peek R. M, Shaw L M, and Kelly C P (2001). *cag+ Helicobacter pylori* induce transactivation of the epidermal growth factor receptor in AGS gastric epithelial cells. *J. Biol. Chem.*, **276**, 48127–48134.

Kim J S, Kim J M, Jung H C, and Song I S (2001). Caspase-3 activity and expression of Bcl-2 family in human neutrophils by *Helicobacter pylori* water-soluble proteins. *Helicobacter*, **6**, 207–215.

Klemke R L, Cai S, Giannini A L, Gallagher P J, de Lanerolle P, and Cheresh D A (1997). Regulation of cell motility by mitogen-activated protein kinase. *J. Cell Biol.*, **137**, 481–492.

Kuipers E J, Perez-Perez G I, Meuwissen S G M, and Blaser M J (1995). *H. pylori* and atrophic gastritis: Importance of the *cag*A status. *J. Natl. Cancer I.*, **87**, 1777–1780.

Kurosawa A, Miwa H, Hirose M, Tsune I, Nagahara A, and Sato N (2002). Inhibition of cell proliferation and induction of apoptosis by *Helicobacter pylori* through increased phosphorylated p53, p21 and Bax expression in endothelial cells. *J. Med. Microbiol.*, **51**, 85–91.

Lee A, O'Rourke J, Corazon de Ungria M, Robertson B, Daskalopoulos G, and Dixon M F (1997). A standardized mouse model of *Helicobacter pylori* infection: Introducing the Sydney strain. *Gastroenterology*, **112**, 1386–1397.

Le'Negrate G, Ricci V, Hofman V, Mograbi B, Hofman P, and Rossi B (2001). Epithelial intestinal cell apoptosis induced by *Helicobacter pylori* depends on expression of the *cag* pathogenicity island phenotype. *Infect. Immun.*, **69**, 5001–5009.

Leung W K, To K F, Chan F K L, Lee T L, Chung S C S, and Sung J J Y (2000). Inter-action of *Helicobacter pylori* eradication and non-steroidal anti-inflammatory drugs on gastric epithelial apoptosis and proliferation: Implications on ul-cerogenesis. *Aliment. Pharm. Therapeut.*, **14**, 879–885.

Li S D, Kersulyte D, Lindley I J D, Neelam B, Berg D E, and Crabtree J E (1999). Multiple genes in the left half of the cag pathogenicity island of *Helicobacter pylori* are required for tyrosine kinase-dependent transcription of interleukin-8 in gastric epithelial cells. *Infect. Immun.*, **67**, 3893–3899.

Liu E S, Wong B C, and Cho C H (2001). Influence of gender difference and gastritis on gastric ulcer formation in rats. *J. Gastroen. Hepatol.*, **16**, 740–747.

Lynch D A, Mapstone N P, Clarke A M, Jackson P, Dixon M F, Quirke P, and Axon A T (1995a). Cell proliferation in the gastric corpus in *Helicobacter pylori* associated gastritis and after gastric resection. *Gut*, **36**, 351–353.

Lynch D A, Mapstone N P, Clarke A M, Sobala G M, Jackson P, Morrison L, Dixon M F, Quirke P, and Axon A T (1995b). Cell proliferation in *Helicobacter pylori* associated gastritis and the effect of eradication therapy. *Gut*, **36**, 346–350.

Lynch D A F, Mapstone N P, Clarke A M T, Jackson P, Moayyedi P, Dixon M F, Quirke P, and Axon A T R (1999). Correlation between epithelial cell pro-liferation and histological grading in gastric mucosa. *J. Clin. Pathol.*, **52**, 367–371.

Machado J C, Pharoah P, Sousa S, Carvalho R, Oliveira C, Figueiredol C, Amorim A, Seruca R, Caldas C, Carneiro F, and Sobrinho-Simoes M (2001). Interleukin-1β and interleukin 1RN polymorphisms are associated with in-creased risk of gastric carcinoma. *Gastroenterology*, **121**, 823–829.

Maeda S, Yoshida H, Mitsuno Y, Hirata Y, Ogura K, Shiratori Y, and Omata M (2002). Analysis of apoptotic and antiapoptotic signalling pathways induced by *Helicobacter pylori*. *Gut*, **50**, 771–778.

Matsumoto K and Nakamura T (1996). Heparin functions as a hepatotrophic factor by inducing production of hepatocyte growth factor. *Biochem. Biophys. Res. Commun.*, **227**, 455–461.

McCarthy C J, Crofford L J, Greenson J, and Scheiman J M (1999). Cyclo-oxygenase-2 expression in gastric antral mucosa before and after eradica-tion of *Helicobacter pylori* infection. *Am. J. Gastroenterol.*, **94**, 1218–1223.

Miehlke S, Yu J, Ebert M, Szokodi D, Vieth M, Kuhlisch E, Buchcik R, Schimmin W, Wehrmann U, Malfertheiner P, Ehninger G, Bayerdorffer E, and Stolte M (2002). Expression of G1 phase cyclins in human gastric cancer and gastric mucosa of first-degree relatives. *Digest. Dis. Sci.*, **47**, 1248–1256.

Mitsuno Y, Maeda S, Yoshida H, Hirata Y, Ogura K, Akanuma M, Kawabe T, Shiratori Y, and Omata M (2002). *Helicobacter pylori* activates the proto-oncogene c-fos through SRE transactivation. *Biochem. Biophys. Res. Com-mun.*, **291**, 868–874.

Miyazaki Y, Shinomura Y, Tsutsui S, Zushi S, Higashimoto Y, Kanayama S, Higashiyama S, Taniguchi N, and Matsuzawa Y (1999). Gastrin induces heparin-binding epidermal growth factor-like growth factor in rat gastric epithelial cells transfected with gastrin receptor. *Gastroenterology*, **116**, 78–89.

Moss S F, Sordillo E M, Abdulla A M, Makarov V, Hanzely Z, Perez-Perez G I, Blaser M J, and Holt P R (2001). Increased gastric epithelial cell apoptosis associated with colonisation with cagA+ *Helicobacter pylori* strains. *Cancer Res.*, **61**, 1406–1411.

Murayama Y, Miyagawa J, Shinomura S, Kanayama S, Isozaki K, Mizuno H, Ishiguro S, Kiyohara T, Miyazaki Y, Taniguchi N, Higahiyama S, and Matsuzawa Y (2002). Significance of the association between heparin-binding epidermal growth factor-like growth factor and CD9 in human gastric cancer. *Int. J. Cancer*, **98**, 505–513.

Naef M, Yokoyama M, Friess H, Buchler M W, and Korc M (1996). Co-expression of heparin-binding EGF-like growth factor and related peptides in human gastric carcinoma. *Int. J. Cancer*, **66**, 315–321.

Naumann M, Wessler S, Bartsch C, Wieland B, Covacci A, Haas R, and Meyer T F (1999). Activation of activator protein 1 and stress response kinases in epithelial cells colonized by *Helicobacter pylori* encoding the pathogenicity island. *J. Biol. Chem.*, **274**, 31655–31662.

Naumann M (2001). Host cell signaling in *Helicobacter pylori* infection. *Int. J. Med. Microbiol.*, **291**, 299–305.

Naumann M and Crabtree J E (2004). *Helicobacter* pylori-induced epithelial cell signalling in gastric carcinogenesis. *Trends Microbiol.*, **12**, 29–36.

Nobes C D and Hall A (1999). Rho GTPases control polarity, protrusion, and adhesion during cell movement. *J. Cell Biol.*, **144**, 1235–1244.

Odenbreit S, Puls J, Sedlmaier B, Gerland E, Fischer W, and Haas R (2000). Translocation of CagA into epithelial cells by type IV secretion. *Science*, **287**, 1497–1500.

Ogura K, Maeda S, Nakao M, Watanabe T, Tada M, Kyutoku T, Yoshida H, Shiratori Y, and Omata M (2000). Virulence factors of *Helicobacter pylori* responsible for gastric diseases in Mongolian gerbil. *J. Exp. Med.*, **192**, 1601–1610.

Panella C, Ierardi E, Polimeno L, Balzano T, Ingrosso M, Amoruso A, Traversa A, and Francavilla A (1996). Proliferative activity of gastric epithelium in progressive stages of *Helicobacter pylori* infection. *Digest. Dis. Sci.*, **41**, 1132–1138.

Peek R M, Moss S F, Tham K T, Perez-Perez G I, Wang S, Miller G G, Atherton J C, Holt P R, and Blaser M J (1997). *Helicobacter pylori cagA*+ strains and dissociation of gastric epithelial cell proliferation and apoptosis. *J. Natl. Cancer I.*, **89**, 863–868.

Peek R M, Blaser M J, Mays D J, Forsyth M H, Cover T L, Song S Y, Krishna U, and Pietenpol J A (1999). *Helicobacter pylori* strain-specific genotypes and modulation of the gastric epithelial cell cycle. *Cancer Res.* **59**, 6124–6131.

Peek R M, Wirth H P, Moss S F, Yang M, Abdalla A M, Tham K T, Zhang T, Tang L H, Modlin I M, and Blaser M J (2000). *Helicobacter pylori* alters gastric epithelial cell cycle events and gastrin secretion in Mongolian gerbils. *Gastroenterology*, **118**, 48–59.

Peek R M and Blaser M J (2002). *Helicobacter pylori* and gastrointestinal tract adenocarcinomas. *Nat. Rev. Cancer*, **2**, 28–37.

Philpott D J, Belaid D, Troubadour P, Thiberge J M, Tankovic J, Labigne A, and Ferrero R L (2002). Reduced activation of inflammatory responses in host cells by mouse-adapted *Helicobacter pylori* isolates. *Cell. Microbiol.*, **4**, 285–296.

Polat A, Cinel L, Dusmez D, Aydin O, and Egilmez R (2002). Expression of cell-cycle related proteins in *Helicobacter pylori* gastritis and association with gastric carcinoma. *Neoplasma*, **49**, 95–100.

Pomorski T, Meyer T F, and Naumann M (2001). *Helicobacter pylori*-induced prostaglandin E_2 synthesis involves activation of cytosolic phospholipase A_2 in epithelial cells. *J. Biol. Chem.*, **276**, 804–810.

Ponzetto C, Bardelli A, Zhen Z, Maina F, dalla Zonca P, Giordano S, Graziani A, Panayotou G, and Comoglio P M (1994). A multifunctional docking site mediates signaling and transformation of the hepatocyte growth factor/scatter factor reception family. *Cell*, **77**, 261–271.

Prenzel N, Zwick E, Daub H, Leserer M, Abraham R, Wallasch C, and Ullrich A (1999). EGF receptor transactivation by G-protein-coupled receptors requires metalloprotease cleavage of proHB-EGF. *Nature*, **402**, 884–88.

Rokkas T, Ladas S, Liatsos C, Petridou E, Papatheodorou G, Theocharis S, Karameris A, and Raptis S (1999). Relationship of *Helicobacter pylori* CagA status to gastric epithelial cell proliferation and apoptosis. *Digest.Dis. Sci.*, **44**, 487–493.

Romano M, Ricci V, Memoli A, Tuccillo C, Di Popolo A, Sommi P, Acquaviva A M, Del Vecchio Blanco C, Bruni C B, and Zarrilli R (1998). *Helicobacter pylori* up-regulates cyclooxygenase-2 mRNA expression and prostaglandidn E_2 synthesis in MKN 28 gastric mucosal cells in vitro. *J. Biol. Chem.*, **273**, 28560–28563.

Romano M, Ricci V, Di Popolo Sommi P, Del Vecchio Blanco C, Bruni C B, Ventura U, Cover T L, Blaser M J, Coffey R J, and Zarrilli R (1998). *Helicobacter pylori* up-regulates expression of epidermal growth factor-related peptides, but inhibits their proliferative effect in MKN-28 gastric mucosal cells. *J. Clin. Invest.*, **101**, 1604–1613.

Roth R A, Kapadia S B, Martin S M, and Lorenz R G (1999). Cellular immune responses are essential for development of *Helicobacter felis* associated gastric pathology. *J. Immunol.*, **163**, 1490–1497.

Rudi J, Kuck D, Strand S, von Herbay A, Mariani S M, Krammer P H, Galle P R, and Stremmel W (1998). Involvement of the CD95 (APO1/Fas) receptor and ligand system in *Helicobacter pylori*-induced gastric epithelial apoptosis. *J. Clin. Invest.*, **102**, 1506–1514.

Sakagami T, Dixon M F, O'Rourke J, Howlett R, Alderuccio F, Vella J, Shimoyama T, and Lee A (1996). Atrophic gastric changes in both *Helicobacter felis* and *Helicobacter pylori* infected mice are host dependent and separate from antral gastritis. *Gut*, **39**, 639–648.

Satoh K, Mutoh H, Eda A, Yanaka I, Osawa H, Honda S, Kawata H, Kihira K, and Sugano K (2002). Aberrant expression of CDX2 in the gastric mucosa with and without intestinal metaplasia: Effect of eradication of *Helicobacter pylori*. *Helicobacter*, **7**, 192–198.

Sawaoka H, Kawano S, Tsuji S, Tsuji M, Sun W, Gunawan E S, and Hori M (1998). *Helicobacter pylori* infection induces cyclooxygenase-2 expression in human gastric mucosa. *Prostag. Leukotr. Ess.*, **59**, 313–316.

Segal E D, Cha J, Falkow S, and Tompkins L S (1999). Altered states: Involvement of phosphorylated CagA in the induction of host cellular growth changes by *Helicobacter pylori*. *Proc. Natl. Acad. Sci. USA*, **96**, 14559–14564.

Sharma S A, Tummuru M K R, Miller G G, and Blaser M J (1995). Interleukin-8 response of gastric epithelial cell lines to *Helicobacter pylori* stimulation in vitro. *Infect. Immun.*, **63**, 1681–1687.

Shen Y, Naujokas M, Park M, and Ireton K (2000). InIB-dependent internalization of *Listeria* is mediated by the Met receptor tyrosine kinase. *Cell*, **103**, 501–510.

Shirin H, Sordillo E M, Oh S H, Yamamoto H, Delohery T, Weinstein I B, and Moss S F (1999). *Helicobacter pylori* inhibits the G1 to S transition in AGS gastric epithelial cells. *Cancer Res.*, **59**, 2277–2281.

Shirin H, Sordillo E M, Kolevska T K, Hibshoosh H, Kawabata Y, Oh S H, Kuebler J F, Delohery T, Weghorst C M, Weinstein I B, and Moss S F (2000). Chronic *Helicobacter pylori* infection induces an apoptosis-resistant phenotype associated with decreased expression of p27(kip1). *Infect. Immun.*, **68**, 5321–5328.

Smythies L E, Waites K B, Lindey J R, Harris P R, Ghiara P, and Smith P D (2000). *Helicobacter pylori* induced mucosal inflammation is Th1 mediated and exacerbated in IL-4 but not IFN γ gene deleted mice. *J. Immunol.*, **165**, 1022–1029.

Sommer F, Faller G, Rollinghoff M, Kirchner T, Mak T W, and Lohoff M (2001). Lack of gastritis and of an adaptive immune response in interferon regulatory factor-1 deficient mice infected with *Helicobacter pylori*. *Eur. J. Immunol.*, **31**, 396–402.

Sommi P, Savio M, Stivala L A, Scotti C, Mignosi P, Prosperi E, Vannini V, and Solcia E (2002). *Helicobacter pylori* releases factor(s) inhibiting cell cycle progression of human gastric cell lines by affecting cyclin E/cdk2 kinase activity and Rb protein phosphorylation through enhanced p27^{KIP1} protein expression. *Exp. Cell Res.*, **281**, 128–139.

Sozzi M, Crosatti M, Kim S K, Romero J, and Blaser M J (2001). Heterogeneity of *Helicobacter pylori cag* genotypes in experimentally infected mice. FEMS *Microbiol. Lett.*, **203**, 109–114.

Stein M, Rappuoli R, and Covacci A (2000). Tyrosine phosphorylation of *Helicobacter pylori* CagA antigen after cag driven host cell translocation. *Proc. Natl. Acad. Sci. USA*, **97**, 1263–1268.

Suerbaum S, Smith J M, Bapurnia K, Morelli G, Smith N H, Kunstmann E, Dyrek I, and Achtman M (1998). Free recombination within *Helicobacter pylori*. *Proc. Natl. Acad. Sci. USA*, **95**, 12619–12624.

Sung J J, Leung W K, Go M Y, To K F, Cheng A S, Ng E K, and Chan F K (2000). Cyclooxygenase-2 expression in *Helicobacter pylori*-associated premalignant and malignant gastric lesions. *Am. J. Pathol.*, **157**, 729–35.

Suzuki M, Raab G, Moses M A, Fernandez C A, and Klagsbrun M (1997). Matrix metalloprotease-3 releases active heparin-binding EGF-like growth factor by cleavage at a specific juxtamembrane site. *J. Biol. Chem.*, **272**, 31730–31737.

Thiery J P (2002). Epithelial-mesenchymal transitions in tumour progression. *Nat. Rev. Cancer*, **2**, 442–454.

Tsutsui S, Shinomura Y, Higashiyama S, Higashimoto Y, Miyasaki Y, Kanayama S, Hiraoka S, Minami T, Kitamura S, Murayama Y, Miyagawa J, Taniguchi N, and Matsuzawa Y (1997). Induction of heparin binding epidermal growth factor-like growth factor and amphiregulin mRNAs by gastrin in the rat stomach. *Biochem. Biophys. Res. Commun.*, **235**, 520–523.

Uemura N, Okamoto S, Yamamoto S, Matsumura N, Yamaguchi S, Yamakido M, Taniyama K, Sasaki N, and Schlemper R J (2001). *H. pylori* infection and the development of gastric cancer. *New Engl. J. Med.*, **345**, 784–789.

van Doorn N E M, Namavar F, Sparrius M, Stoof J, van Rees E P, van Doorn L J, and Vandenbrouke-Grauls C M (1999). *Helicobacter pylori*-associated gastritis in mice is host and strain specific. *Infect. Immun.*, **67**, 3040–3046.

Varro A, Noble P J, Wroblewski L E, Bishop L, and Dockray G J (2002). Gastrin-cholecystokininB receptor expression in AGS cells is associated with direct inhibition and indirect stimulation of cell proliferation via paracrine activation of the epidermal growth factor receptor. *Gut*, **50**, 827–833.

Wallasch C, Crabtree J E, Bevac D, Robinson P A, Wagner H, and Ullrich A (2002). *Helicobacter pylori* stimulated EGF receptor transactivation requires metalloprotease cleavage of HB-EGF. *Biochem. Biophys. Res. Commun.*, **295**, 695–701.

Wang J, Fan X, Lindholm C, Bennett M, O'Connoll J, Shanahan F, Brooks E G, Reyes V E, and Ernst P B (2000a). *Helicobacter pylori* modulates lymphoepithelial cell interactions leading to epithelial cell damage through Fas/Fas Ligand interactions. *Infect. Immun.*, **68**, 4303–4311.

Wang J, Court M, Jeremy A H T, Aboshkiwa M A, Robinson P A, and Crabtree J E (2003). Infection of Mongolian gerbils with Chinese *Helicobacter pylori* strains. *FEMS Immunol. Med. Mic.*, **36**, 207–213.

Wang T C, Koh T J, Varro A, Cahill R J, Dangler C A, Fox J G, and Dockray G J (1996). Processing and proliferative effects of human progastrin in transgenic mice. *J. Clin. Invest.*, **98**, 1918–1929.

Wang T C, Goldenring J R, Dangler C, Ito S, Mueller A, Jeon W K, Koh T J, and Fox J G (1998). Mice lacking secretory phospholipase A2 show altered apoptosis and differentiation with *Helicobacter felis* infection. *Gastroenterology*, **114**, 675–89.

Wang T C, Dangler C A, Chen D, Goldenring J R, Koh T, Raychowdhury R, Coffey R J, Ito S, Varro A, Dockray G J, and Fox JG (2000b). Synergistic interaction between hypergastrinemia and *Helicobacter* infection in a mouse model of gastric cancer. *Gastroenterology*, **118**, 36–47.

Watanabe T, Tada M, Nagai H, Sasaki S, and Nakao M (1998). *Helicobacter pylori* infection induces gastric cancer in Mongolian Gerbils. *Gastroenterology*, **115**, 642–642.

Webb P M, Crabtree J E, and Forman D (1999). Gastric cancer, cytotoxin-associated gene A positive *H. pylori* and serum pepsinogens: An international study. *Gastroenterology*, **116**, 269–276.

Weel J F L, Hulst R W M van der, Gerrits Y, Roorda P, Feller M, Dankert J, Tytgat G N J, and Van der Ende A (1996). The interrelationship between cytotoxin-associated gene A, vacuolating cytotoxin, and *Helicobacter pylori*-related diseases. *J. Infect. Dis.*, **173**, 1171–1175.

Weidner K M, Di Cesare S, Sachs M, Brinkmann V, Behrens J, and Birchmeier W (1996). Interaction between Gab1 and the c-Met receptor tyrosine kinase is responsible for epithelial morphogenesis. *Nature*, **384**, 173–176.

Wessler S, Hocker M, Fischer W, Wang T C, Rosewicz S, Haas R, Wiedenmann B, Meyer T F, and Naumann M (2000). *Helicobacter pylori* activates the histidine decarboxylase promoter through a mitogen-activated protein kinase pathway independent of pathogenicity island-encoded virulence factors. *J. Biol. Chem.*, **275**, 3629–3636.

Wessler S, Rapp U R, Wiedenmann B, Meyer T F, Schoneberg T, Hocker M, and Naumann M (2002). B-Raf/Rap1 signaling, but not c-Raf-1/Ras, induces the histidine decarboxylase promoter in *Helicobacter pylori* infection. *FASEB J.*, **16**, 417–419.

Wirth H P, Beins M H, Yang M, Tham K T, and Blaser M J (1998). Experimental infection of Mongolian gerbils with wild-type and mutant *Helicobacter pylori* strains. *Infect. Immun.*, **66**, 4856–4866.

Wong B C, Wang W P, So W H, Shin V Y, Wong W M, Fung F M, Liu E S, Hiu W M, Lam S K, and Cho C H (2001). Epidermal growth factor and its receptor in chronic active gastritis and gastroduodenal ulcer before and after *Helicobacter pylori* eradication. *Aliment. Pharm. Therap.*, **15**, 1459–1465.

Xia H H and Talley N J (2001). Apoptosis in gastric epithelium induced by *Helicobacter pylori* infection: Implications in gastric carcinogenesis. *Am. J. Gastroenterol.*, **96**, 16–26.

Xie H, Pallero M A, Gupta K, Chang P, Ware M F, Witke W, Kwiatkowski D J, Lauffenburger D A, Murphy-Ullrich J E, and Wells A (1998). EGF receptor regulation of cell motility: EGF induces disassembly of focal adhesions independently of the motility-associated PLCγ signaling pathway. *J. Cell Sci.*, **111**, 615–624.

Yoshimura T, Tomita T, Dixon M F, Axon A T R, Robinson P A, and Crabtree J E (2002). ADAMs (A Disintegrin and Metalloproteinase) messenger RNA expression in *H. pylori*-infected, normal and neoplastic gastric mucosa. *J. Infect. Dis.*, **185**, 332–340.

Yu J, Russell R M, Saloman R N, Murphy J C, Palley L S, and Fox J G (1995). Effect of *Helicobacter mustelae* infection on ferret gastric epithelial cell proliferation. *Carcinogenesis*, **16**, 1927–1931.

Yu J, Leung W K, Ng E K, To K F, Ebert M P, Go M Y, Chan W Y, Chan F K, Chung S C, Malfertheiner P, and Sung J J (2001). Effect of *Helicobacter pylori* eradication on expression of cyclin D2 and p27 in gastric intestinal metaplasia. *Aliment. Pharm. Therapeut.*, **15**, 1505–1511.

Yu J, Leung W K, Go M Y Y, Chan M C W, To K F, Ng E K W, Chan F K L, Ling T K W, Chung S C S, and Sung J J Y (2002). Relationship between *Helicobacter pylori* babA2 status with gastric epithelial cell turnover and premalignant lesions. *Gut*, **51**, 480–484.

CHAPTER 9

Bacteria and cancer

Christine P J Caygill and Michael J Hill[1]

It has been estimated that more than 80% of human cancers are caused by environmental factors, and so are potentially preventable (Higginson, 1968). Bacteria are in a unique position to mediate between the host and its external environment because they naturally colonise all external surfaces. In addition, bacterial infections expose the host to further interactions with bacteria.

The role of the flora in protecting the host against infection has been well documented. The normal bacterial flora of the mouth protects against fungal infection; the normal flora of the intestine helps to protect against enteritis; the normal skin flora helps to protect against skin infection.

Furthermore, the gut bacterial flora has an overwhelmingly beneficial role in assisting digestion and in detoxifying ingested compounds of exogenous origin (Hill, 1995). However, there are situations where bacterial infection, which can be defined as bacteria being in the wrong place, can produce carcinogens and is associated with local cancers (e.g., bladder infection, nitrosamines, and bladder cancer). In addition, there are situations where the normal flora, carrying out its normal metabolic activity, can be exposed to very high levels of a substrate resulting in toxic levels of the product (e.g., colon bacteria, bile acids, and colon cancer).

In this chapter we will first discuss the types of carcinogens/mutagens that can be produced by bacterial action, and the situations in which this production might be significant. In addition, the gut bacteria release a wide range of anti-carcinogens from the diet as well, and these are discussed. In the final section the possible contribution of bacteria to the risk of cancer at specific sites, and the mechanism by which that might occur, are discussed.

---

[1] Deceased, to whom this article is dedicated.

THE MECHANISMS INVOLVED IN CANCER

A full description of the molecular mechanisms involved in carcinogenesis is beyond the scope of this chapter, but a brief description will be helpful for understanding how bacteria could impinge on the carcinogenic process. Much has been learned in the past 20 to 30 years that provides a general insight into carcinogenesis.

It is now clear that mutation of key cellular genes enables the cell to escape the normal control checkpoints that regulate entry into the cell cycle and cellular growth. An integrated network of signalling pathways interpret signals, both soluble and from cell–cell interactions, arriving at cell surface receptors. The net result is to control gene expression and activation of proteins involved in cell cycle regulation. Many of these signalling proteins, cell cycle regulation proteins, and proteins involved in programmed cell death (apoptosis) are oncogenic when mutated, and are often found in mutant form in human cancers.

There are various stages to carcinogenesis: initiation, whereby a mutation, either spontaneous or induced, in a relevant gene begins the process of loss of normal growth regulation; promotion, where this aberrant cell grows to produce a colony; and progression, where the aberrant cell acquires more mutations, becomes more deregulated, and is able to grow outside its normal cellular environment. These processes do not necessarily occur in this order, as increased growth (i.e., promotion) can provide an increased opportunity for mutation, (i.e., initiation) to occur.

Bacterial factors could impinge on this process either by the production of factors that are mutagenic and cause initiation, or by promoting growth in a number of ways (Lax and Thomas, 2002).

BACTERIAL METABOLISM AND CARCINOGENS/ANTICARCINOGENS

Interest in the ability of bacteria to produce carcinogens derived much of its impetus from the hypothesis that large bowel cancer was caused by a bacterial metabolite produced *in situ* in the colon from some benign substrate (Aries et al., 1969; Hill et al., 1971). In examining that hypothesis, a very wide range of metabolites was identified that were carcinogens/mutagens/modulators of carcinogenesis. These have been summarised by Hill (1986), and will be described briefly below.

At the time of that review, most interest was in the possible factors that caused cancer. However, later it was realised that the production of anticarcinogens might be much more important, and so the second half of this section concerns those compounds.

CARCINOGENS/MUTAGENS/PROMOTERS

Microbes produce carcinogens/mutagens/tumour promoters by three main routes, namely (1) *de novo*, (2) by metabolism of exogenous substrates, or (3) by release from their glycoside conjugates.

De Novo Synthesis

The carcinogens/mutagens produced *de novo* by microbes that are best characterised are the mycotoxins to which humans may be exposed as food contaminants or as therapeutic agents.

The best example is the aflatoxin (Wogan, 1973), produced by *Aspergillus* spp, which commonly contaminate foods (such as peanuts, beans, or maize) that are improperly stored. They are a particular problem in tropical areas where the climate is very suitable for fungal survival and growth. They have been shown to be potent carcinogens in a wide range of animal species, where they normally cause hepatocellular carcinomas. In humans, there is strong epidemiological evidence relating aflatoxin exposure in Hepatitis B virus carriers to hepatic carcinoma risk in a number of African and Asian countries (see section on liver cancer).

Fusarium toxins are produced by a wide range of field fungi belonging to *Fusarium spp.* (Schoental, 1979). These toxins have been shown to be gastrointestinal carcinogens in a number of animal species, and have been implicated (Schoental, 1979) in human gastrointestinal tract cancers. However, there have been no long-term follow-up studies of patients known to be exposed to fusarium toxin.

There are a number of carcinogenic mycotoxins that are useful anticancer therapeutic agents (e.g., mitomycin C, adriamycin) or antimicrobial agents (e.g., actinomycin, azaserine, griseofulvin). The chemotherapeutic agents are used only for late-stage cancers, in whom the risk of any long-term carcinogenicity is unimportant.

Production by Metabolism of Exogenous or Endogenous Substrates

Some of the products of bacterial metabolism of exogenous substrates with carcinogenic or mutagenic properties are listed in Table 9.1.

Table 9.1 *Carcinogens/promoters produced by gut bacteria from dietary substrates*

Substrate	Product	Action
Protein		
• Tyrosine	Volatile phenols	Promoters
• Tryptophan	Various metabolites	
• Basic amino acids	Cyclic secondary amines	
	τ N-nitroso compounds	
• Methionine	ethionine	Carcinogens
Fat		
• Lecithin	Dimethylamine τ N-nitroso	Carcinogen
• Primary bile salts	Secondary bile acids	Promoters
Carbohydrate		
• Sugars	Fermentation to butyrate	
• Lignin	Lignans	Protection
Glycosides		
• Plant glycosides	Flavonoids	Protective
	Polyphenolic compounds	
	Tannins	
• Cycasin	Anthocyanins	Carcinogen
	Methylazoxymethanol	

Amino Acid Metabolites

Carcinogenic metabolites can be metabolised from a number of amino acids including tryptophan, tyrosine, and methionine (Hill, 1986; 1995). Tryptophan can be metabolised by bacteria to yield a range of products that have been associated with the risk of human bladder cancer (Bryan, 1971). Tyrosine is metabolised to a range of volatile phenols that are excreted in the urine, but have been shown to be tumour promoters in rodents. Methionine is metabolised to its *S*-ethyl analogue, ethionine, which is a potent liver carcinogen in animals. Furthermore, the basic amino acids lysine, arginine, and ornithine can be metabolised to cyclic secondary amines which can be nitrosated to produce highly carcinogenic *N*-nitrosamines

N-Nitroso Compounds (NNC)

The NNC are a class of potent carcinogens that have proved to be carcinogenic in all of the animal species in which they have been tested. In consequence, although there is no unambiguous proof that they are carcinogenic in humans, there is no good reason to believe that humans are uniquely resistant. The

Table 9.2 *Sites in the human body where bacterial synthesis of N Nitroso compounds (NNC) has been demonstrated. For references see Hill (1996)*

Body site	Comments
Saliva	*In vitro* incubation
Normal Acid stomach	Acid catalysed reaction; salivary nitrite and endogenous nitrogen compounds
Achlorhydric stomach	Bacterially catalysed formation from dietary nitrate and endogenous nitrogen compounds
Infected urinary bladder	Infecting organisms catalyses NNC formation from urinary substrates
Bilharzia	Mixed bacterial infection catalyses formation of much greater amounts of NNC
Colon in urine diversion	Profuse mixed bacterial population of the colon, and the high levels of nitrate and nitrosatable amine in urine result in very high NNC production.
Infected cervix	Nitrate is secreted into the cervix; which is a rich source of amines and has a rich mixed bacterial flora.

carcinogenicity of the NNC has been reviewed regularly (e.g., Magee, 1996); the NNC can be divided into *N*-nitrosamines and *N*-nitrosamides (which includes *N*-nitrosureas and *N*-nitrosguanidine). Whereas these latter are powerful locally acting carcinogens, the *N*-nitrosamines are target-organ specific, the target varying between animal species (Magee and Barnes, 1967).

NNC are formed by the action of nitrous acid on a suitable secondary nitrogen compound. This reaction can occur chemically with an optimum pH of 2, and can also be catalysed by bacterial enzymes at neutral pH values. NNC can therefore be formed anywhere in the body where bacteria, nitrate or nitrite, and a suitable nitrosatable amine occur. Their production has been demonstrated (Hill, 1996) in saliva, gastric juice, infected urine, the colon, and the infected cervix (Table 9.2), and they have been implicated in local carcinogenesis at most of those sites.

Bile Acid Metabolites

Bile acids play an important role in fat digestion. The bile acids synthesised by the liver are cholic and chenodeoxycholic acids (both of which are nontoxic to the mucosa), and these are secreted in the bile as their glycine or taurine conjugates. These bile conjugates then act to emulsify dietary fat and aid its absorption from the small bowel. They remain in the lumen of the small bowel but are absorbed by an active transport mechanism from the terminal ileum and returned to the liver. This entero-hepatic circulation is a very efficient recovery system, with only 1–5% of the pool being lost to the colon with each circulation. However, on a high fat diet there is an increased rate of entero-hepatic circulation of the bile acid pool, and about 800 mg of bile acid enters the colon each day of which about 500 mg is lost per day in faeces.

The bile conjugates undergo extensive metabolism by the gut bacteria in the colon; the principal reactions are deconjugation to release the free bile acids (which can then be passively absorbed from the colon), and 7-dehydroxylation to release deoxycholic and lithocholic acids from cholic and chenodeoxycholic acids respectively. There is a considerable body of literature to demonstrate that these two "secondary bile acids" can be mutagenic, co-mutagenic, or tumour promoting (Hill, 1986).

Release of Carcinogens/Mutagens from Their Glycoside Conjugates

Two groups of substrates are of interest under this heading. The first is the glucuronide conjugates of polycyclic aromatic hydrocarbon (PAH) carcinogens, and the second is the plant glucosides.

There is continuous environmental exposure to hydrocarbon carcinogens such as the PAH. These are detoxified by the liver by hydroxylation, then conjugated with glucuronic acid and secreted in bile for excretion. These conjugates are hydrolysed by bacterial beta glucuronidase, and although the hydroxy-PAH aglycone is non-carcinogenic, there is a carcinogenic high-energy intermediate formed during the hydrolysis step (Kinoshita and Gelboin, 1978).

A wide range of plant products with pharmacological activity are present in the form of mainly harmless inactive beta glucosides. These include the cathartic glucosides (e.g., senna, cascara), the cardioactive glucosides (e.g., lanatoside C), the cyanogenetic glucosides (e.g., amygdalin) and the carcinogenic glucosides (e.g., cycasin). Cycasin is non-toxic in germ-free rodents; it is the beta glucoside of methylazoxymethanol, and the aglycone, released by bacterial beta glucosidase, is highly hepatotoxic and carcinogenic. The carcinogenicity of cycasin was reviewed by Laqueur and Spatz (1968), and cycasin is widely used as a colon carcinogen in animal models.

There is a wide range in the activity of the two glycosidases as produced by different genera of gut bacteria (Table 9.3). However, because of their

Table 9.3 *Activity of β-glucuronidase and β-glucosidase produced
by various genera of gut bacteria (Hill, 1986)*

| Organisms | Enzyme activity (moles substrate degraded per hour per 10^8 cells; mean of 50 strains) | |
	β-glucuronidase	β-glucosidase
Escherichia coli	24.7 ± 2.1	5.8 ± 2.5
Streptococcus faecalis	2.9 ± 0.6	192.7 ± 19.5
Lactobacillus spp	1.6 ± 0.2	26.0 ± 7.4
Clostridium spp	11.3 ± 2.3	22.1 ± 5.0
Bacteroides spp	6.0 ± 3.5	35.1 ± 4.8
Bifidobacterium spp	1.9 ± 0.8	29.3 ± 6.0

numerical predominance in the gut flora, in practice only the enzyme activities of the *Bacteroides spp* and the *Bifidobacterium spp* contribute *in vivo*.

ANTI-CARCINOGENS

Although in early studies attention was focussed almost exclusively on the
production or release of carcinogens by bacterial action, the production
or release of anti-carcinogens may be equally important. Plants produce a
wide range of anutrients with anti-carcinogenic properties (Johnson et al.,
1994). These include flavonoids, flavonols, isoflavones, polyphenolics, tan-
nins, isothiocyanates, etc. Most of these are present in the plant in the glucoside form, and so would need the action of bacterial beta glucosidase to
release the active component.

Intakes of most of the plant anti-carcinogens have yet to be correlated
with any particular cancer site. However, the phytoestrogens, a group of
naturally occurring compounds with estrogenic properties found in plant
foods, have been implicated in protection against breast and prostate cancer.
They can be divided into a number of classes, the principal of which are the
isoflavones, the coumestans, and the lignans. The phytoestrogens include
genistein, which is known to interfere with cellular signalling by inhibiting
tyrosine phosphorylation of proteins.

The lignans are a class of phenolic compounds related to the polyphenolic
plant structural compounds, the lignins. Whereas the lignins are insoluble
very high molecular weight structural polymers, the lignans are relatively
low molecular weight and highly water soluble. The isoflavones are virtually

found only in legumes (and particularly soya), but the lignans are found in almost all cereals and vegetables, with the highest concentrations found in oilseeds. Like many plant anutrients of interest, they are usually present in plant foods as their glycosides. Like the related lignins, the lignans have no nutritive value to humans and until the last 25 years were largely ignored. Many have been shown to have an anti-mitotic action. Most importantly for this chapter, the lignans in human urine are not aglycones of plant lignans but are produced in the colon by bacterial action on dietary precursors, then absorbed from the colon and excreted in the urine (Adlercreutz, 1984). The urinary concentrations can be greatly decreased by the action of antibiotics. Insertion of a biliary fistula showed that the lignans undergo enterohepatic circulation. Their concentration in urine is increased with increasing intake of dietary fibre and of whole-grain foods, and decreased by increasing tobacco consumption. They are weak estrogens, able to bind to oestrogen binding sites, particularly the ER-beta sites.

There has been a growing interest in the study of these compounds in recent years because consumption of foods rich in phytoestrogens has been vigorously advocated for the prevention of breast, prostate, and colon cancer. The incidence of all three cancers is lower in Asia than in Western countries; the intakes of phytoestrogens in general (and lignans in particular) is much higher in Asian populations than in those living in the West (Adlercreutz, 1998, 2002). The principal lignans in human urine are enterodiol and enterolactone; the concentration of both is higher in the urine of Asian populations (with a low incidence of the cancers of interest) than of European populations. The intake of the main food sources of lignans, the whole grain cereals, is inversely related to risk of cancer at a number of sites, including the breast, prostate, and colon (La Vecchia and Chatenoud. 1998), although other lifestyle factors could also be responsible for the differences.

BACTERIAL INFECTIONS AND CANCER

Cancers of the Urinary Tract

Bladder cancer has long been known to be associated with industrial exposures to naphthylamines, benzidine, and a range of aromatic amines. This was the reason why the vast majority of cases arose in males in industrialised countries. However, a proportion of cases is not of industrial origin. Early anecdotal evidence suggested an excess risk of bladder cancer following chronic bladder infection. Radomski et al. (1978) confirmed an association between chronic simple urinary tract infection and subsequent bladder cancer. Bladder infections are very common, and often asymptomatic (Sinclair and Tuxford, 1971); the data on cancer risk reported by Radomski et al. (1978) concerns chronic symptomatic infection resistant to therapy, but many of his controls might have had asymptomatic bladder infections and so the magnitude of the excess risk would have been underestimated.

There is copious evidence that carcinogenic N-nitroso compounds (NNC) are produced *in situ* in the bladder by infecting organisms; this is to be expected because the urine is the route of excretion of the substrates for NNC production – nitrate and nitrosatable amines. Radomski et al. (1978) suggested that these NNC were the cause of the cancer.

Bilharzial infection is a major risk factor for bladder cancer, and such infections are accompanied by a profuse secondary bacterial infection of the bladder. Hicks et al. (1977) showed strong evidence that the bladder cancer associated with bilharzial infection was due to the NNC produced by the secondary bacterial infection. They also produced evidence that the excess risk of bladder cancer in paraplegia was due to the same mechanism – NNC produced by a chronic bacterial infection of the bladder.

Recently it has been shown that what were thought to be recurrent bladder infections by *Escherichia coli* are in fact chronic infections (Mulvey et al., 2001). The majority of bacterial isolates from the urinary tract are *E. coli*, including from cases of prostatitis (Andreu et al., 1997). Most such isolates express the cytotoxic necrotizing toxin (CNF) that stimulates the important signalling protein Rho (see Chapter 3). CNF has been shown to be important for pathogenesis in the urinary tract (Rippere-Lampe et al., 2001a; 2001b); the presence of CNF can lead to activation of the enzyme Cyclooxygenase-2 (COX-2) (Thomas et al., 2001). Numerous studies have implicated COX-2 in both the initiation and progression of various cancers (Liu et al., 2001).

It has also been shown that a history of urinary tract infection increases the risk of renal cancers (Parker et al., 2004), so it is possible that several different types of urinary tract cancers could have in part a bacterial cause.

Gastric Cancer

In 1975 Correa et al. proposed a hypothesis for the sequence of events that led from the normal to the neoplastic stomach (Figure 9.1). Although this sequence has since been added to and changed the essential hypothesis remains the same (Figure 9.1), and there are two types of bacterial contamination that may play a role; one is infection with *Helicobacter pylori* and the other is chronic bacterial overgrowth as a result of hypochlohydria of the stomach.

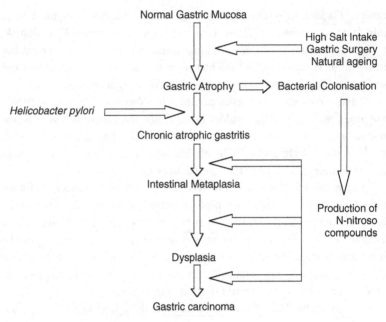

Normal Gastric Mucosa

High Salt Intake
Gastric Surgery
Natural ageing

Gastric Atrophy ⟹ Bacterial Colonisation

Helicobacter pylori ⟹

Chronic atrophic gastritis

Intestinal Metaplasia

Production of
N-nitroso
compounds

Dysplasia

Gastric carcinoma

Figure 9.1. The pathogenesis of gastric cancer.

Helicobacter pylori Infection

H. pylori infection is associated with low socioeconomic status and crowded living conditions, especially in childhood (Malaty and Graham, 1994). World-wide, about half the population are infected (Smith and Parsonnet, 1998), with most children in developing countries being infected by the age of 10. In contrast, in developed countries, infection in children is uncommon and only 40–50% of adults are affected. There is a clear age-related increase in prevalence that is probably due to a cohort effect in that *H. pylori* infection in childhood was more common in the past than it is today (Parsonnet et al., 1992; Banatvala et al., 1993). The route of transmission of *H. pylori* remains controversial, with circumstantial evidence suggesting that it probably occurs through person-to-person transmission.

There have been a number of studies comparing rates of *H. pylori* infection in different populations with rates of gastric cancer in the same populations (Forman et al., 1990; Eurogast Study Group, 1993). These have mostly correlated well, as has the decline in gastric cancer rate over time with the decline in *H. pylori* incidence (Parsonnet et al., 1992; Banatvala et al., 1993).

As cancerous stomachs may lose their ability to harbour *H. pylori* (Osawa et al., 1996), retrospective studies have to be viewed with caution. However,

CHRISTINE P J CAYGILL AND MICHAEL J HILL

two meta-analyses of all case-control studies (Huang et al., 1998; Eslick and Talley, 1998) indicate a 2-fold increase in the risk of gastric cancer in instances of *H. pylori* infection.

More concrete evidence for a link has come from prospective case-control studies in which stored serum from populations was used and infection was known to precede malignancy (Parsonnet et al., 1991; Forman et al., 1991; Lin et al., 1995; Siman et al., 1997). It has also been shown that in those followed up for more than 10 years the risk of gastric cancer was increased 8-fold in those infected with *H. pylori* (Forman et al., 1994).

A more detailed discussion of the potential mechanisms involved in *H. pylori* carcinogenesis is given in Chapter 8 by Naumann and Crabtree.

Chronic Bacterial Overgrowth of the Stomach

Chronic bacterial overgrowth of the stomach occurs when the normally acid pH of the stomach (pH 2) rises on a permanent basis to pH 4.5 or above. This occurs as part of the ageing process, but also in a number of pathological conditions, such as pernicious anaemia (PA), and in people who have had surgery for peptic ulcer.

PA is caused by a lack of intrinsic factor, which is accompanied by a failure to secrete gastric acid. Indeed, hypochlorhydria is a symptom in the recognition and diagnosis of the disease.

Surgical treatment of peptic ulcer was directed towards decreasing gastric acid secretion, whether by gastrectomy, where most of the lower part of the stomach was removed by a variety of procedures, or vagotomy, where the vagal nerves which control acid secretion were severed. Each of these procedures resulted in loss of gastric acidity within a year, and in each case there was an increased risk of gastric cancer (Caygill et al., 1984).

A number of authors (Blackburn et al., 1968; Brinton et al., 1989) have reported an increased risk of gastric cancer in PA patients. In our study (Caygill et al., 1990), we found a 5-fold excess risk of gastric cancer in PA patients, but it was not possible to establish an accurate latency period as the onset of PA could not be ascertained accurately for a sufficient number. However, we were able to divide up the period after diagnosis into 0–19 years and 20+ years and found that the excess risk of gastric cancer was 4-fold in the first time period and 11-fold in the second time period.

There have been a number of studies showing an increased risk of gastric cancer in those who have had a gastrectomy for benign disease, and these are summarised in Table 9.4 (from the review by Caygill and Hill, 1992). In our own study (Caygill et al., 1986), we analysed cancer risk by time interval for those who had had a gastrectomy for gastric ulcer (GU) separately from

Table 9.4 *Gastric cancer following surgery for peptic ulcer*

Reference:	Size of study population	Type of Study	Excess risk & latency/ follow-up period
Stalsberg and Taksdal (1971)	630	Case-control	4-fold after 15 years
Ross et al. (1982)	779	Cohort	None over 19 years
Sandler et al. (1984)	521	Case-control	None over 20-year follow-up
Watt et al. (1984)	735	Cohort	3-fold after 15 years over a 15–25 year follow-up
Tokudome et al. (1984)	3827	Cohort	None over 10–30 years follow-up
Caygill et al. (1986)	4466	Cohort	4-fold after 20 years
Viste et al. (1986)	3470	Cohort	3-fold after 20 years
Arnthorsson et al. (1988)	1795	Cohort	2-fold after 15 years
Lundegardh et al. (1988)	6459	Cohort	3-fold after 30 years
Toftgaard (1989)	4131	Cohort	2-fold after 25 years
Offerhaus et al. (1988)	2633	Cohort	5-fold after 15 years in females, 3 fold after 25 years in males
Caygill et al. (1991)	1643	Cohort	1.6-fold over 20 years

those who had the operation for duodenal ulcer (DU). We found that in the case of DU there was a decrease in risk in the first 19 years, followed by an increase in risk thereafter. In contrast, in the GU patients there was a 3-fold increase in risk immediately after, and presumably prior to surgery, and this rose to over 5-fold 20 or more years after surgery. The pattern of an initial decrease in risk in those operated for DU has been confirmed by Arnthorsson et al. (1988), Moller and Toftgaard (1991), Lundegardh (1988) and Eide et al. (1991). This difference in behaviour between DU and GU patients needs to be rationalised. Prior to surgery DU patients have good acid secretion and the effect of surgery is to render them hypoacidic. Many GU patients on the other hand are often hypoacidic for a number of years prior to the operation.

The histopathological sequence (Figure 9.1) from the normal to the neoplastic stomach, suggested and reviewed by Correa (1988), has been generally accepted. Correa et al. (1975) postulated that the first stage, gastric atrophy, progresses to chronic atrophic gastritis. This carries an increased risk of development of intestinal metaplasia, with consequent increased risk of increasingly severe dysplasia and finally carcinoma. Gastric atrophy results in the loss of gastric acid secretion, which allows bacterial proliferation. The bacteria then react with nitrate, present in many foods and also in drinking water, to convert it to nitrite, which in turn is converted to carcinogenic N-nitroso compounds. The latter potent carcinogens were postulated as the cause of the progression through intestinal metaplasia and increasingly severe dysplasia to cancer. If this hypothesis is correct, then the loss of gastric acidity, with consequent chronic bacterial overgrowth from any cause (surgical, metabolic, clinical, genetic, or environmental) should, after a latency period of 20 years or more, lead to an increased risk of gastric cancer. Indeed, this has also been shown to be the case in PA patients, post-gastrectomy patients, and those undergoing vagotomy (Caygill et al., 1991). The above hypothesis explains the difference in cancer risk in those operated on for a GU and a DU. As a result of their hypoacidity prior to operation, the GU patients will have had bacterial overgrowth for variable lengths of time that would contribute to the latency period, whereas those with DU would have become hypochlorhydric only after their operation and the increase in their risk would start to manifest itself only 20 years later.

Colorectal Cancer

Large bowel carcinogenesis is a multistage process with at least three distinct histological stages (Hill and Morson, 1978; Hill, 1991; Hill et al., 2001). These are (a) adenoma formation, (b) adenoma growth, and (c) increasingly severe dysplasia to malignancy and potential or actual metastatic disease. The evidence for the dysplasia–carcinoma sequence (formerly known as the adenoma–carcinoma sequence) was reviewed by Morson (1974) and by Morson et al. (1983). The steps in this pathway are distinct (Figure 9.2), and have different controlling factors.

Adenoma formation is an extremely common event in Western populations, and postmortem studies show their prevalence to be approximately 50% in men and 30% in women by age 70. The vast majority of these are tiny, and are presumably not seen at endoscopy. They remain tiny and asymptomatic and therefore are normally detected only at postmortem. The risk of finding a malignant component in a tiny adenoma is very low (less that 1 per 1000 for adenomas less than 3 mm diameter), whereas it is high in adenomas greater than 20 mm diameter (Morson et al., 1983). Clearly, therefore, adenoma growth is an important step on the adenoma–carcinoma sequence.

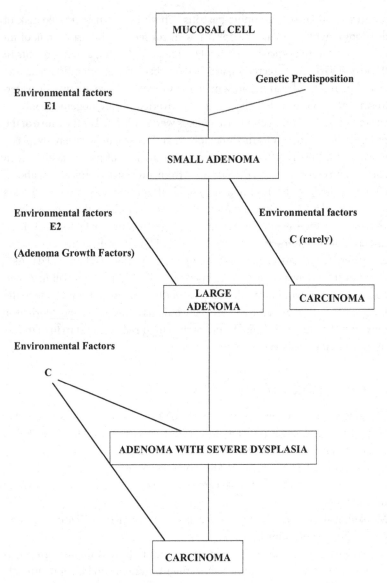

Figure 9.2. The hypothesised mechanism of colorectal carcinogenesis based on that proposed by Hill and Morson (1978).

There are differences in epidemiology between small adenomas, large adenomas, and colorectal cancers. One of the most crucial differences is the differing subsite distribution. In a very large number of postmortem studies (Hill, 1986), small adenomas have been shown to be evenly distributed around the colorectum, suggesting that the causal agents are delivered via the vascular system, or have a genetic cause. In contrast, large adenomas and cancers are concentrated in the distal colon and rectum. This is consistent with, but does not prove, the hypothesis that the factors causing adenomas to increase in size and in severity of epithelial dysplasia are luminal products of bacterial metabolism; this view is supported by the fact that adenomas regress after diversion of the faecal stream.

The colon lumen is a rich source of potential carcinogens, produced *in situ* by bacterial action on benign substrates (Table 9.1). We were not able to find evidence associating the risk of colon cancer with the faecal concentrations of the amino acid metabolites, and the best evidence suggests a role for secondary bile acids (Table 9.5) in the adenoma growth stage (Hill, 1991). This evidence comes from comparisons of populations with colon cancer risk, case control studies, and studies of high-risk patient groups.

There were two major groups of high-risk patients. The adenoma patients were part of a follow-up study, and were followed for 10 years. There was no relation between bile acids and the rate of formation of new adenomas, showing that bile acids have no role in adenoma formation. However there was a good correlation with the size of the largest adenoma, supporting the hypothesis that bile acids are implicated in adenoma growth. The colitics had all been patients for more than 10 years and had been offered a colectomy. They opted instead to join a follow-up clinic, and were followed for a further 10 years. During that time dysplasia (analogous to adenoma formation) was detected in 44 out of 112 patients. Although there was no relation between bile acid concentration and formation of dysplasia per se, there was a good correlation with the severity of dysplasia.

In addition, there is a mass of supporting evidence from animal model studies (Reddy, 1992). The evidence that bile acids are tumour promoters or mutagens or co-mutagens is summarised in Table 9.2.

Gallbladder Cancer

Cancer of the gallbladder is rare in Western countries, but is most frequent in the Andean countries of South and Central America and in American Indian groups (Devor, 1982) and in some parts of India (Shukla et al., 2000). It has a poor prognosis. The aetiology is not well understood, but known risk

Table 9.5 *The evidence implicating faecal bile acid (FBA) concentration in colorectal cancer (CRC) risk (Hill, 1991)*

Study type	Observation
1. Comparison of populations	In ten studies of populations, the FBA concentration correlated with CRC risk
2 Case-control studies	In some, but not all, case control studies, the FBA concentration was higher in CRC cases than in controls.
3. Diet studies	Diet factors that decrease risk of CRC (e.g., cereal fibre) also decrease FBA concentration
4. Animal diet studies	Diet manipulations that increase CRC risk increase FBA concentration and vice versa.
5. Patient groups	Surgical treatments that increase CRC risk (e.g., partial gastrectomy, cholecystectomy) also increase FBA concentration.
6. Bile acid binding sites	Bile acid binding sites were detected in 31% of CRC patients and less than 2% of controls.
7. *In vitro* studies	Bile acids are tumour promoters in animal models and co-mutagenic in microbial mutagenesis assays.
8. Mucosal toxicity	Bile acids cause dysplastic changes in the rodent colon.

factors are gallstones, Polya partial gastrectomy, and chronic infection with *Salmonella typhi/paratyphi* for peptic ulcer. One feature common to all these risk factors is the association with bacterial infection.

Devor (1982) reviewed 69 reports of series of gallbladder cancer cases. Of these 59 reports (4184 cases) had details of gallstone status, and of these 77% were associated with gallstone carriage. The nature of this association is not clear, but it is known that gallstones are associated with bacterial infection of the gallbladder (England and Rosenblatt, 1977).

The aim of Polya partial gastrectomy for peptic ulcer disease is to decrease acid secretion into the stomach. This is achieved by removal of most of the lower part (including much of the acid-secreting section) of the stomach and results in the stomach becoming hyperchlorhydric with a pH of about 4.5. This is a perfect milieu for bacterial overgrowth and formation of *N*-Nitroso compounds (see p. 202), which have been shown to be carcinogenic in all species in which they have been studied. Polya partial gastrectomy is associated with a 10-fold excess risk of gallbladder cancer with a 20-year latency period (Caygill et al., 1988).

There is a growing body of evidence that typhoid carriers are at an increased risk of biliary tract cancer. The largest study reported to date is a case-control study of 471 carriers registered by the New York City Health Department and 942 age- and sex-matched controls. It showed that chronic carriers were six times as liable to die of hepatobiliary cancer as controls (Welton, 1979) and this has been confirmed by others (Mellemgaard and Gaarslev, 1988; Caygill et al., 1994; Nath et al., 1997).

We investigated the long-term cancer risk in two cohorts – one a cohort of 386 acute typhoid cases from a single outbreak and the other 83 typhoid carriers from a number of different outbreaks (Caygill et al., 1994; Caygill et al., 1995). Tables 9.6 and 9.7 show cancer risk in the two cohorts. In the case of acute infection in the 1964 Aberdeen outbreak, there was no excess risk from cancer of the gallbladder, nor indeed from any other cancer, but in the cohort with chronic infection there was an almost 200-fold excess risk of cancer of the gallbladder.

There is no doubt that gallbladder cancer has a multifactorial aetiology, which may well differ according to the individual's environmental exposure. One important aspect of this appears to be exposure to chronic bacterial infection of the gallbladder.

Oesophageal Cancer

Although *H. pylori* infection is accepted as a risk factor in gastric cancer, its role in oesophageal cancer is less well established. Siman et al. (2001) found that infection with *H. pylori* was associated with a decreased risk of oesophageal malignancy, the protective effect being more pronounced for oesophageal adenocarcinoma (odds ratio 0.16) than for squamous cell carcinoma (odds ratio 0.41). In a case control study (22 resection patients with adenocarcinoma compared to 22 age- and sex-matched resection patients with squamous cell carcinoma), *H. pylori* was seen in one adenocarcinoma case and five squamous cell cancer controls (Cameron et al., 2002).

Adenocarcinoma of the oesophagus results from persistent gastroesophageal reflux disease, particularly Barrett's oesophagus, and an increase in this condition in the United States and Europe is concomitant with a decline in the prevalence of infection with *H. pylori* in these populations. This and the effect of therapy are discussed in two reviews by Sharma (2001) and Koop (2002).

(216)

Table 9.6 *Death from cancer and from "all causes" in a cohort of typhoid/paratyphoid cases from the 1964 Aberdeen outbreak*

Cause of death	ICD no.	O	E		O/E		95% CI	
Biliary tract	156	0	0.23	(0.23)	0	(0)	0–16.0	(0–16.0)
Pancreas	157	0	1.47	(1.46)	0	(0)	0–2.51	(0–2.53)
Colorectum	152–4	3	4.40	(4.09)	0.68	(0.73)	0.14–1.99	(0.15–2.14)
Stomach	151	2	2.18	(1.86)	0.92	(1.08)	0.11–3.31	(0.13–3.88)
Lung	162	6	8.55	(6.84)	0.70	(0.88)	0.26–1.53	(0.32–1.91)
All neoplasms	140–208	23	32.55	(29.00)	0.71*	(0.79)	0.45–1.06	(0.50–1.19)
Death from all causes		81	141.68	(126.93)	0.55**	(0.64)**	0.45–0.71	(0.51–0.79)

ICD, International Classification of Diseases; O, observed; E, expected; CI, confidence interval.

* $P < 0.05$; ** $P < 0.001$.

Expected mortality using Grampian Area statistics are in parentheses.

Table 9.7 *Deaths from cancer and from "all causes" in the typhoid/paratyphoid carrier cohort*

Cause of Death	ICD no.	Male				Female				Total			
		O	E	O/E	95% CI	O	E	O/E	95% CI	O	E	O/E	95% CI
Gallbladder	1560	0	<0.005	0		5	0.03	167	(54–389)	5	0.03	167***	(54–391)
Pancreas	157	3	0.13	23.1***	(4.8–67.4)	0	0.24	0	(0–15.4)	3	0.37	8.1***	(1.7–23.7)
Colorectum	152–4	0	0.32	0	(0–11.5)	3	0.68	4.4*	(0.9–12.9)	3	1.00	3.8	(0.6–8.8)
Lung	162	0	1.10	0	(0–3.35)	5	0.88	5.7**	(1.8–13.3)	5	1.98	2.5	(0.8–5.9)
All neoplasms	140–208	6	2.94	2.0	(0.7–4.4)	14	4.86	2.9***	(1.6–4.8)	20	7.80	2.6***	(1.6–4.0)
All causes		13	13.64	1.0	(0.5–1.6)	45	24.60	1.8***	(1.3–2.5)	58	38.24	1.5**	(1.2–2.0)

ICD, International Classification of Diseases; O, observed; E, expected; CI, confidence interval.

$*P < 0.05$; $**P < 0.01$; $***P < 0.001$.

Expected mortality using Grampian Area statistics are in parentheses.

Lung Cancer

There have been numerous reviews of the association between tuberculosis (TB) and subsequent lung cancer. Before 1950, most TB patients died at a young age, before their risk of lung cancer became manifest. In consequence, therefore, the co-existence of the two diseases was rare. Indeed, as a result of early studies there was a theory (Rokitansky, 1854) that the two diseases were antagonistic. The association was not manifest, therefore, until TB treatment regimes were sufficiently successful to give the patient a reasonable life expectancy. Aoki (1993) reviewed the epidemiological studies between 1960 and 1990 and confirmed that patients with active pulmonary tuberculosis have an excess risk of dying of lung cancer, despite their already high mortality from tuberculosis. The excess was 5–10 fold depending on age, and was greater in women that in men. Patients with active disease were the most likely to develop lung cancer, but he also found that TB cases had excess risks of cancers at other sites (e.g., colon, lymphoma, myeloma).

The mechanism for the association is not clear. It has been shown that mycobacterial products stimulate protein kinase C (Sueoka et al., 1995), a signalling molecule activated by the tumour-promoting phorbol esters, and this might contribute to carcinogenesis. In animal studies, attempts to stimulate the immune surveillance system with BCG resulted in increased, rather than decreased, cancer risks (Martin et al., 1977), and this could be a reason for the increased risk of cancer at distant sites seen by Aoki (1993).

Pancreatic Cancer

The two circumstances where cancer of the pancreas is linked with bacterial infection are typhoid carriage and 20 years after surgery for peptic ulcer.

N-Nitroso compounds can be formed in any situation where nitrite and nitrosatable amines are present together. Surgery for peptic ulcer results in gastric hypochlorhydria with resultant bacterial overgrowth. These bacteria react with ingested nitrates in food and convert them to nitrites. These are highly reactive compounds that can combine with nitrosatable amines, also present in food, and form a range of nitrosamines (see p. 202). Nitrosamines have been found to be carcinogenic in a number of animal species (Magee and Barnes, 1967). They are both species- and target-organ–specific and there is no reason to believe that humans are uniquely immune from their action. This could well be the explanation for the finding of an excess risk for cancer of the pancreas after surgery for peptic ulcer, the excess risk being greater in GU than in DU patients (Caygill et al., 1987; Mack et al., 1986; Eide et al., 1991; Tersmette et al., 1990; Ross et al., 1982). It must be noted, however, that others found no excess risk in patients operated on for peptic ulcer (Moller and Toftgaard, 1991; Inokuchi et al., 1984; Watt et al., 1984).

Caygill et al. (1994) found a large excess (23-fold) in cancer of the pancreas in male typhoid carriers, though not in acute cases of typhoid who did not become carriers. The mechanism is uncertain, but pancreatic cancer has been associated with bile reflux from the common bile duct (Wynder et al., 1975).

CONCLUSIONS

The role for bacteria in human carcinogenesis is still largely hypothetical, although some bacteria have clearly been shown to promote carcinogenesis, most notably *H. pylori*. However, the molecular mechanisms involved in bacterially induced cancer remain largely unclear. For a convincing proof of the association between bacteria and cancer, it would be necessary to show all of the following:

(1) bacteria are able to produce metabolites, or specific products, that can influence either the initiation of carcinogenesis, or the rate of its progression to cancer; and

(2) bacteria actually produce these metabolites *in vivo*; and

(3) epidemiology indicates that there is an association between the risk of the cancer and the amounts of bacterial metabolite that is not coincidental and could be causal.

There is clear and indisputable evidence for the first two factors. There is good evidence for the last one, which is stronger for some sites (e.g., bladder) than for others. If the hypothesis could be proved for a particular cancer site, it could have important consequences for the prevention of such cancers. Although this topic has been controversial for over a century, the recent increased interest in this area suggests that a more profound understanding of the interactions involved in bacterially induced cancer will become apparent within the next few years.

REFERENCES

Adlercreutz H (1984). Does fiber-rich food containing animal lignan precursers protect against both colon and breast cancer? An extension of the "'fiber hypothesis.'" *Gastroenterology*, **86**, 761–766.

Adlercreutz H (1998). Human health and phytoestrogens. In *Reproductive and Developmental Toxicology*, ed. K S Korach, pp. 299–371, Marcel Dekker, New York.

Adlercreutz H (2002). Phytoestrogens and cancer. *Lancet Oncol.*, **3**, 364–373.

Andreu A, Stapleton A E, Fennell C, Lockman H A, Xercavins M, Fernandez F, and Stamm W (1997). Urovirulence determinants in *Escherichia coli* strains causing prostatitis *J. Infect. Dis.*, **176**, 464–469.

Aoki K (1993). Excess incidence of lung cancer among pulmonary tuberculosis cases. *Jpn. J. Clin. Oncol.*, **23**, 205–220.

Aries V C, Crowther J S, Drasar B S, Hill M J, and Williams R E (1969). Bacteria and aetiology of cancer of large bowel. *Gut*, **10**, 334–335.

Arnthorsson G, Tulinuis H, Egilsson V, Sigvaldason H, Magnusson B, and Thorarinsson H (1988). Gastric-cancer after gastrectomy. *Int. J. Cancer*, **42**, 365–367.

Banatvala N, Mayo K, Megraud F, Jennings R, Deeks J J, and Feldman R A (1993). The cohort effect and *Helicobacter pylori*. *J. Infect. Dis.*, **168**, 219–221.

Blackburn E K, Callende S T, Dacie J V, Doll R, Girdwood R H, Mollin D L, Saracci R, Stafford J L, Thompson R B, Varadi S, and Wetherle G (1968). Possible association between pernicious anaemia and leukaemia: A prospective study of 1,625 patients with a note on very high incidence of stomach cancer. *Int. J. Cancer*, **3**, 163–170.

Brinton L A, Gridley G, Hrubec Z, Hoover R, and Fraumeni J F (1989). Cancer risk following pernicious-anaemia. *Brit. J. Cancer*, **59**, 810–813.

Bryan G T (1971). Role of urinary tryptophan metabolites in etiology of bladder cancer. *Am. J. Clin. Nutr.*, **24**, 841–847.

Cameron A J, Souto E O, and Smyrk T C (2002). Small adenocarcinomas of the esophagogastric junction: Association with intestinal metaplasia and dysplasia. *Am. J. Gastroenterol.*, **97**, 1375–80.

Caygill C, Hill M, Craven J, Hall R, and Miller C (1984). Relevance of gastric achlorhydria to human carcinogenesis. In: *N-nitroso Compounds: Occurrence, Biological Effects and Relevance to Human Cancer*, ed. I K O'Neil, R C Van Borstel, C T Miller, J Long, and H Bartsch, pp. 895–900, Lyon: IARC (Scientific Publication No 57).

Caygill C, Hill M, Kirkham J, and Northfield T C (1988). Increased risk of biliary-tract cancer following gastric-surgery. *Brit. J. Cancer*, **57**, 434–436.

Caygill C P J, Braddick M, Hill M J, Knowles R L, and Sharp J C M (1995). The association between typhoid carriage, typhoid infection and subsequent cancer at a number of sites. *Eur. J. Cancer Prev.*, **4**, 187–193.

Caygill C P J and Hill M J (1992). Malignancy following surgery for benign peptic disease – A review. *Ital. J. Gastroenterol.*, **24**, 218–224.

Caygill C P J, Hill M J, Braddick M, and Sharp J C M (1994). Cancer mortality in chronic typhoid and paratyphoid carriers. *Lancet* **343**, 83–84.

Caygill C P J, Hill M J, Kirkham J S, and Northfield T C (1986). Mortality from gastric-cancer following gastric surgery for peptic-ulcer. *Lancet*, **I**, 1505–1505.

Caygill C P J, Knowles R L, and Hall R (1991). Increased risk of cancer mortality after vagotomy for peptic ulcer: A preliminary analysis. *Eur. J. Cancer Prev.*, 1, 35–37.

Caygill C P J, Knowles R L, and Hill M J (1990). The relationship between pernicious anaemia and gastric cancer. *Deut. Zeit. Für Onkologie*, 22, 120–122.

Caygill C P J, Hill M J, Hall C N, Kirkham J S, and Northfield T C (1987). Increased risk of cancer at multiple sites after gastric-surgery for peptic-ulcer. *Gut*, 28, 924–928.

Correa P, Haenszel W, Cuello C, Tannenbaum S, and Archer M (1975). Model for gastric cancer epidemiology. *Lancet*, 2, 58–60.

Correa P (1988). Precancerous lesions of the stomach phenotypic changes and their determinants. In *Gastric Carcinogenesis*, ed. P I Reed and M J Hill, pp .127–136, Amsterdam, New York, Oxford: Exerpta Medica.

Devor E J (1982). Ethnogeographic patterns in gallbladder cancer. In *Epidemiology of Cancer of the Digestive Tract*, ed. P Correa and W Haenszel, pp. 197–225, Martinus Nijhof, The Hague.

Eide T J, Viste A, Andersen A, and Soreide O (1991). The risk of cancer at all sites following gastric operation for benign disease – a cohort of 4224 patients. *Int. J. Cancer*, 48, 333–339.

England D M and Rosenblatt J E (1977). Anaerobes in human biliary tracts. *J. Clin. Microbiol.*, 6, 494–500.

Eslick G D and Talley N J (1998). *Helicobacter pylori* infection and gastric carcinoma: A meta-analysis. *Gastroenterology* 114, G2428 Part 2 Suppl. S.

Eurogast Study Group: Forman D, Debacker G, Elder J, Moller H, Damotta L C, Roy P, Abid L, Debacker G, Tjonneland A, Boeing H, Haubrich T, Wahrendorf J, Manousos O, Kafatos A, Tulinius H, Ogmundsdottir H, Palli D, Cipriani F, Fukao A, Tsugane S, Miyajima Y, Zatonski W, Tycznski J, Calheiros J, Zakelj M P, Potocnik M, Webb P, Knight T, Wilson A, Kaye S, Potter J, Newell D G, Hengels K J, Kyrtopoulos S, Wild C, Moller H, Webb P, Newell D, Forman D, Palli D, Hengels K, Elder J, Debacker G, and Coleman M (1993). Epidemiology of, and risk-factors for, *Helicobacter pylori* infection among 3194 asymptomatic subjects in 17 populations. *Gut*, 34, 1672–1676.

Forman D, Newell D G, Fullerton F, Yarnell J W G, Stacey A R, Wald N, and Sitas F (1991). Association between infection with *Helicobacter pylori* and risk of gastric-cancer – evidence from a prospective investigation. *Brit. Med. J.*, 302, 1302–1305.

Forman D, Sitas F, Newell D G, Stacey A R, Boreham J, Peto R, Campbell T C, Li J, and Chen J (1990). Geographic association of *Helicobacter-pylori* antibody prevalence and gastric-cancer mortality in rural China. *Int. J. Cancer*, 46, 608–611.

Forman D, Webb P, and Parsonnet J (1994). *Helicobacter-pylori* and gastric-cancer. *Lancet*, **343**, 243–244.

Hicks R M, Walters C L, Elsebai I, Elaasser A B, Elmerzabani M, and Gough T A (1977). Demonstration of nitrosamines in human urine – Preliminary observations on a possible etiology for bladder cancer in association with chronic urinary-tract infections. *P. Roy. Soc. Med.*, **70**, 413–416.

Higginson J (1968). Theoretical possibilities of cancer prevention in man. *P. Roy. Soc. Med.*, **61**, 723–726.

Hill M J (1986). *Microbes and Human Carcinogenesis*. Edward Arnold, London.

Hill M J (1991). Bile acids and colorectal cancer: Hypothesis. *Eur. J. Cancer Prev.*, **1** (suppl 2), 69–74.

Hill M J (1995). Role of gut bacteria in human toxicology and pharmacology. Taylor and Francis, London.

Hill M J (1996). Endogenous N-nitrosation. *Eur. J. Cancer Prev.*, **5** (suppl 1), 47–50.

Hill M J, Crowther J S, Drasar B S, Hawkswor G, Aries V C, and Williams R E (1971). Bacteria and aetiology of cancer of large bowel. Lancet **I**, 95–100.

Hill M J and Morson B C (1978). Etiology of adenoma-carcinoma sequence in large bowel. Lancet **I**, 245–7.

Hill M J, Davies G J, and Giacosa A (2001). Should we change our dietary advice on cancer prevention? *Eur. J. Cancer Prev.*, **10**, 1–6.

Huang J, Sridhar S, Chen Y, and Hunt R H (1998). Meta-analysis of the relationship between *Helicobacter pylori* seropositivity and gastric cancer. *Gastroenterlogy*, **114**, 1169–1179.

Inokuchi K, Tokudome S, Ikeda M, Kuratsune M, Ichimiya H, Kaibara N, Ikejiri T, and Oka N (1984). Mortality from carcinoma after partial gastrectomy. *Gann*, **75**, 588–594.

Johnson I T, Williamson G, and Musk S R R (1994). Anticarcinogenic factors in plant foods: A new class of nutrients? *Nutr. Res. Rev.*, **7**, 175–204.

Kinoshita N and Gelboin H (1978). Beta-glucuronidase catalyzed-hydrolysis of benzo[a]pyrene-3-glucuronide and binding to DNA. *Science*, **199**, 307–309.

Koop H (2002). Gastroesophageal reflux disease and Barrett's esophagus. *Endoscopy*, **34**, 97–103.

La Vecchia C and Chatenoud L (1998). Fibres, whole-grain foods and breast and other cancers. *Eur. J. Cancer Prev.*, **7** (Suppl 2), S25–S28.

Laqueur G L and Spatz M (1968). Toxicology of cycasin. *Cancer Res.*, **28**, 2262–2267.

Lax A J and Thomas W (2002). How bacteria could cause cancer: One step at a time. *Trends Microbiol.*, **10**, 293–299.

Lin J T, Wang L Y, Wang J T, Wang T H, Yang C S, and Chen C J (1995). A nested case-control study on the association between *Helicobacter-pylori* infection and gastric-cancer risk in a cohort of 9775 men in Taiwan. *Anticancer Res.*, **15**, 603–606.

Liu C H, Chang S H, Narko K, Trifan O C, Wu M T, Smith E, Haudenschild C, Lane T F, and Hla T (2001). Over expression of cyclooxygenase-2 is sufficient to induce tumorigenesis in transgenic mice. *J. Biol. Chem.*, **276**, 18563–18569.

Lundegardh G, Adami H O, Helmick C, Zack M, and Meirik O (1988). Stomach-cancer after partial gastrectomy for benign ulcer disease. *New. Engl. J. Med.*, **319**, 195–200.

Mack T M, Yu M C, Hanisch R, and Henderson B E (1986). Pancreas cancer and smoking, beverage consumption, and past medical history. *J. Natl. Cancer I.*, **76**, 49–60.

Magee P N (1996). Nitrosamines and human cancer: Introduction and overview. *Eur. J. Cancer Prev.*, **5** (*suppl 1*), 7–10.

Magee P N and Barnes J M (1967). Carcinogenic N-nitroso compounds. *Adv. Cancer Res.*, **10**, 163–246.

Malaty H M, Graham D Y (1994). Importance of childhood socioeconomic status on the current prevalence of *Helicobacter-pylori* infection. *Gut*, **35**, 742–745.

Martin M S, Martin F, Justrabo E, Michel M F, and Lagneau A (1977). Immuno-prophylaxis and therapy of grafted rat colonic carcinoma. *Gut*, **18**, 232–235.

Mellemgaard A and Gaarslev K (1988). Risk of hepatobiliary cancer in carriers of *Salmonella typhi*. *J. Natl. Cancer I.*, **80**, 288.

Moller H and Toftgaard C (1991). Cancer occurrence in a cohort of patient surgi-cally treated for peptic ulcer. *Gut*, **32**, 740–744.

Morson B C (1974). Evolution of cancer of colon and rectum. *Cancer* **34**, 845–849.

Morson BC, Bussey H J R, Day D W, and Hill M J (1983). Adenomas of large bowel. *Cancer Surv.*, **2**, 451–478.

Mulvey M A, Schilling J D, and Hultgren S J (2001). Establishment of a persistent *Escherichia coli* reservoir during the acute phase of a bladder infection. *Infect. Immun.*, **69**, 4572–4579.

Nath G, Singh H, and Shukla V K (1997). Chronic typhoid carriage and carcinoma of the gallbladder. *Eur. J. Cancer Prev.*, **6**, 557–559.

Offerhaus G J A, Tersmette A C, Huibregtse K, Vandestadt J, Tersmette K W F, Stijnen T, Hoedemaeker P J, Vandenbroucke J P, and Tytgat G N J (1988). Mortality caused by stomach cancer after remote partial gastrectomy for benign conditions: 40 years of follow-up of an Amsterdam cohort of 2633 postgastrectomy patients. *Gut*, **29**, 1588–1590.

Osawa H, Inoue F, and Yoshida Y (1996). Inverse relation of serum *Helicobacter pylori* antibody titres and extent of intestinal metaplasia. *J. Clin. Pathol.*, **49**, 112–115.

Parker A S, Cerhan J R, Lynch C F, Leibovich B C, and Cantor K P (2004). History of urinary tract infection and risk of renal cell carcinoma. *Am. J. Epidemiol.*, **159**, 42–48.

Parsonnet J, Blaser M J, Perez-Perez G I, Hargrett-Bean N, and Tauxe R V (1992). Symptoms and risk factors of *Helicobacter pylori* infection in a cohort of epidemiolgists. *Gastroenterology*, **102**, 41–46.

Parsonnet J, Friedman G D, Vandersteen D P, Chang Y, Vogelman J H, Orentreich N, and Sibley R K (1991). *Helicobacter-pylori* infection and the risk of gastric-carcinoma. *New Engl. J. Med.*, **325**, 1127–1131.

Radomski J L, Greenwald D, Hearn W L, Block N L, and Woods F M (1978). Nitrosamine formation in bladder infections and its role in the etiology of bladder cancer. *J. Urology*, **120**, 48–56.

Reddy B S (1992). Dietary-fat and colon cancer – animal model studies. *Lipids*, **27**, 807–813.

Rippere-Lampe K E, Lang M, Ceri H, Olson M, Lockman H A, and O'Brien A D (2001a). Cytotoxic Necrotizing Factor type 1-positive *Escherichia coli* causes increased inflammation and tissue damage to the prostate in a rat prostatitis model. *Infect. Immun.*, **69**, 6515–6519.

Rippere-Lampe K E, O'Brien A D, Conran R, and Lockman H A (2001b). Mutation of the gene encoding cytotoxic necrotizing factor type 1 (cnf1) attenuates the virulence of uropathogenic *Escherichia coli*. *Infect. Immun.*, **69**, 3954–3964.

Rokitansky C (1854). *Manual of Pathological Anatomy*. Sydenham Soc, London.

Ross A H M, Smith M A, Anderson J R, and Small W P (1982). Late mortality after surgery for peptic-ulcer. *New Engl. J. Med.*, **307**, 519–522.

Sandler R S, Johnson M D, and Holland K L (1984). Risk of stomach-cancer after gastric-surgery for benign conditions – a case control study. *Digest Dis. Sci.*, **29**, 703–708.

Schoental R (1979). The role of Fusarium mycotoxins in the aetiology of tumours of the digestive tract of man and animals. *Front. Gastr. Res.*, **4**, 17–24.

Sharma P (2001). *Helicobacter pylori*: A debated factor in gastroesophageal reflux disease. *Digest. Dis.*, **19**, 127–33.

Shukla V K, Singh H, Pandey M, Upadhyaya S K, and Nath G (2000). Carcinoma of the gallbladder: Is it a sequel of typhoid? *Digest. Dis. Sci.*, **45**, 900–903.

Siman J H, Forsgren A, Berglund G, and Floren C H (2001). *Helicobacter pylori* infection is associated with a decreased risk of developing oesophageal neoplasms. *Helicobacter*, **6**, 310–6.

Siman J H, Forsgren A, Berglund G, and Floren C H (1997). Association between *Helicobacter pylori* and gastric carcinoma in the city of Malmo, Sweden – A prospective study. *Scand. J. Gastroentero.*, **32**, 1215–1221.

CHRISTINE P J CAYGILL AND MICHAEL J HILL

Sinclair T and Tuxford A F (1971). Incidence of urinary tract infection and asymptomatic bacteriuria in a semi-rural practice. *Practitioner*, **207**, 81–90.

Smith K L and Parsonnet J (1998). *Helicobacter pylori* In *Bacterial Infections of Humans*, ed. A Evans and P Brachman, pp. 337–353, Plenum Publishing, New York.

Stalsberg H and Taksdal S (1971). Stomach cancer following gastric surgery for benign conditions. *Lancet*, **2**, 1175–1177.

Sueoka E, Nishiwaki S, Okabe S, Iida N, Suganuma M, Yano I, Aoki K, and Fujiki H (1995). Activation of protein kinase C by cord factor trenalose 6-monomy colate, resulting in tumor necrosis Factor-alpha release in mouselung tissue, *Jpn. J. Cancer Res.*, **86**, 749–.

Tersmette A C, Offerhaus G J A, Giardiello F M, Tersmette K W F, Vandenbrouke J P, and Tytgat G N J (1990). Occurrence of non-gastric cancer in the digestive tract after remote partial gastrectomy – analysis of an Amsterdam cohort. *Int. J. Cancer*, **46**, 792–795.

Thomas W, Ascott Z K, Harmey D, Slice L W, Rozengurt E, and Lax A J (2001). Cytotoxic Necrotizing Factor from *Escherichia coli* induces RhoA-dependent expression of the cyclooxygenase-2 gene. *Infect. Immun.*, **69**, 6839–6845.

Toftgaard C (1989). Gastric-cancer after peptic-ulcer surgery – A historic prospective cohort investigation. *Ann. Surg.*, **210**, 159–164.

Tokudome S, Kono S, Ikeda M, Kuratsune M, Sano C, Inokuchi K, Kodama Y, Ichimiya H, Nakayama F, Kaibara N, Koga S, Yamada H, Ikejiri T, Oka N, and Tsurumaru H (1984). A prospective study on primary gastric stump cancer following partial gastrectomy for benign gastroduodenal diseases. *Cancer Res.*, **44**, 2208–2212.

Viste P, Opheim P, Thunold J, Eide G E, Bjornestad E, Skarstein A, Hartveit F, Eide T J, and Soreide O (1986). Risk of carcinoma following gastric operations for benign disease – a historical cohort study of 3470 patients. *Lancet*, **2**, 502–505.

Watt P C H, Patterson C C, and Kennedy T L (1984). Late mortality after vagotomy and drainage for duodenal ulcer. *Brit. Med. J.*, **288**, 1335–1338.

Welton J C, Marr J S, and Friedman S M (1979). Association between hepatobiliary cancer and typhoid carrier state. *Lancet* **I**, 791–794.

Wogan G N (1973). Aflatoxin carcinogenesis. In *Methods in Cancer Research*, ed. H Bartsch, p. 309, Academic Press, New York.

Wynder E L (1975). Epidemiological evaluation of causes of cancer of pancreas. *Cancer Res.*, **35**, 2228–2233.

What is there still to learn about bacterial toxins?

Alistair J Lax

The concept that bacteria are rather simple organisms that interact with eukaryotic cells in a passive manner is now totally untenable, as more evidence emerges of active interaction and reaction between bacteria and host. It is also becoming clear that bacterial toxins do not merely operate as molecules of death for a cell. Toxin action may ultimately result in cell death, and indeed death of the host organism, but often the producer bacterium requires that the host cell is first organised in a particular manner. In these circumstances, the toxins sent out by the bacterium on its "behalf" have a mission to regulate the target cell in a very precise manner. Not much is known about the eukaryotic mechanisms that exist to control rather then kill bacteria, but such mechanisms must exist to regulate the homeostatic balance between a eukaryotic host and its commensal bacteria. Furthermore, a host weakened by diseases leading to immunodeficiency, such as HIV/AIDS, stress, or malnutrition, is more susceptible to infection – both by bacteria normally classified as pathogenic, but also by commensal bacteria, which in the circumstances are named as opportunistic pathogens.

Much more is known about the mechanisms that bacteria use to regulate cellular function. Many of these mechanisms have been described in detail in the chapters in this book. Several common themes are appearing. The most frequent targets for such toxins are intracellular proteins involved in the signal transduction pathways that integrate incoming signals at the cell surface. These signalling systems normally lead to an output in terms of cell death, differentiation, or cell growth and division.

Many of these interactions involve proteins of the Rho family. The central role of the Rho protein family in controlling cell shape, motility, and uptake mechanisms has clearly made it attractive for the evolution of bacterial pathogenic mechanisms. Rho proteins are also involved in transmission of signals, both from soluble growth factors that interact with the cell via G-protein coupled receptors and receptor tyrosine kinases, and from cell–cell contact mechanisms. Thus, toxins that attack Rho can produce many different cellular effects. While conventional toxins that target Rho either activate or inactivate its function, some type III delivered effectors act to mimic the function of normal cellular regulators such as Rho GEFs. These latter toxins transiently activate signalling and thus can induce more subtle subversion of signalling.

Some toxins have recently been shown to affect regulation of the cell cycle and the linked process of apoptosis. In this case, however, much less is known about the molecular mechanisms involved.

Despite over a century of work on bacterial toxins, in particular those encoded by the well-known major pathogens, new toxins are still being identified. Indeed, new toxins are still being discovered from well-characterised bacteria, such as the recently described Cif effector from *E. coli* (Marchès et al., 2003). There remains a wealth of possibilities for the discovery of novel toxic activities, particularly from bacteria not viewed as mainstream pathogens or from bacteria not yet identified. In addition, homology searching of sequence data will identify toxins related to known toxins. As more completed bacterial genome sequences become available, homology searching is likely to lead to further identification of toxins.

While many of these "new" toxins are likely to fall into existing categories of toxin action, it is also likely that new mechanisms will be discovered, both in terms of targets and the induced chemical modification. In a similar manner to the effector toxins found to act non-covalently on Rho, other bacterial effectors may act on other signalling components to mimic the action of natural eukaryotic effector molecules. Likewise, it is to be expected that other toxins will be identified that act as mitogens, using either a similar set of molecular targets as the *Pasteurella multocida* toxin (PMT), or entirely different ones. There are also likely to be more toxins that target the machinery of the cell cycle. Such toxins are likely to be valuable tools for the analysis of cell function.

Although the thrust of much recent research has been on the molecular mode of action of toxins, there is clearly a need to understand toxin action *in vivo*. Some toxins have been suggested to play no part in pathogenesis. One example is the dermonecrotic toxin of *Bordetella* species. Mutants in this toxin were shown to display an identical LD_{50} to wild-type bacteria in a murine challenge model (Weiss and Goodwin, 1989). However, this toxin is highly conserved across all the *Bordetella* species and is regulated by the *bvg* His-Asp phosphorelay system, and it thus seems unlikely that the bacteria would have maintained the gene in this functional state if it did not confer an advantage. One recurrent difficulty is the choice of an animal system that is relevant to the human disease, and it may be that DNT has cellular effects that cannot be detected in a LD$_{50}$ test.

It is also possible that toxins will have distinctive effects on different cell types. For example, the stimulation of cellular signalling induced by PMT leads to mitogenicity in some cell types, but its primary effect on osteoblasts is to inhibit differentiation. Thus, although the molecular action of a toxin will be identical between cell types, the manner in which a particular cell integrates the induced signalling can lead to different cellular outcomes.

As has been alluded to in various chapters in this book, several toxins display properties that would suggest that they might operate as tumour promoters. Both PMT and the *E. coli* toxin CNF exhibit such properties. At first sight, it would appear that the multinucleation and perturbation of the cell cycle induced by CNF are unlikely to result in viable cells that could become transformed. However, a potentially important *in vivo* aspect of toxins that has been largely ignored is that not all cells will be exposed to the same level of toxin *in vivo*. While some cells may experience a high concentration of toxin, of the level frequently used experimentally *in vitro*, other cells may be subjected to lower sub-maximal quantities of toxin. Such lower levels would be expected to be less cytotoxic, but might still induce aberrant signalling and could lead to long-term sequelae, such as tumour promotion. This is surely an area that warrants futher attention.

Thus, we can safely predict that toxin research still has a considerable future and that further novel revelations regarding bacterial toxin action will be forthcoming in the years ahead.

REFERENCES

Marchès O, Ledger T N, Boury M, Ohara M, Tu X L, Goffaux F, Mainil J, Rosenshine I, Sugai M, De Ryke J, and Oswald E (2003). Enteropathogenic and enterohaemorrhagic *Escherichia coli* deliver a novel effector called Cif, which blocks cell cycle G2/M transition. *Mol. Microbiol.*, **50**, 1553–1567.

Weiss A A and Goodwin M S (1989). Lethal infection by *Bordetella pertussis* mutants in the infant mouse model. *Infect. Immun.*, **57**, 3757–3764.

Index

AB toxins, 63
  CNF, 39
  CNF, DNT, 43
Actin
  Actin ring and bone resorption, 151
  ADP-ribosylation by C2, 68
  Bartonella and, 92
  Calponin, 35
  Cytoskeleton and Rho family, 33–34
  Myosin light chain, 15
  PMT and, 15
  Rho and, 14
Acting
  Activation by YpkA, 131
Actinobacillus actinomycetemcomitans
  CDT, 59, 64
ADAM metalloprotease disintegrin, 178
  Expression in gastric cancer, 179
Adenoma
  Colorectal cancer and adenoma formation,
    211–213
ADP ribosylation
  Activity of ExoS, 126
  By C2, 68
Aflatoxin, 201
  Hepatitis B virus carriers and liver
    carcinoma, 201
AIDS
  Bartonella in AIDS patients, 83
Akt/PKB
  Activation by RANKL, 150

Alkaline phosphatase, 149
  Down regulation by PMT, 157
Anchorage-independent growth
  PMT and, 17
Angiogenesis
  Bartonella, effects on, 83–84
Animal models
  Carcinogens, 204
  Helicobacter, 171–172, 174–175
Anticarcinogens
  Bacterial, 205–206
AP-1
  Activation by *Bartonella*, 98
Apoptosis
  CNF, 45
  Helicobacter induced, 171
  Helicobacter LPS induced, 174
  PMT and, 20
  Shigella IpaB induced, 134
  Toxin involvement in, 7
  Type III effectors and, 132
Arp2/3
  WASP, 35
Atrophic rhinitis, 155

Bacillary angiomatosis, 82
Bacteria
  Bladder cancer and, 206–207
  Carcinogen production, 200–205
  Stomach cancer linked to bacterial
    overgrowth, 209–211

Bacteroides
  Glycosidase activity, 205
Bartonella, 81–108
  Apoptosis, 106–107
  Cell binding, 84–85
  Cell motility, 99–100
  Disease caused by, 82–83
  Disease in cats, 82
  Effects on angiogenesis, 83–84
  Endothelial cells and, 90–92
  Erythrocytes and, 86–90
  Life cycle, 82
  Perinuclear location, 97
  Porins, 85–86
  Proliferation, 106–107
  Rho and, 92–95, 96, 98
  TFSS, 95
  Upregulation of E-selectin,
    108
  Uptake, 95–96
  VEGF, 104
  VEGF similarity, 100–101, 106
Bile acid metabolites
  Cancer linked to, 203–204
Bilharzial infection
  Bladder cancer and, 207
Bladder cancer, 206
Blebs
  Bartonella, 86
Bombesin, 12, 13
Bone, 147–162
  Dynamics, 147, 151
  Remodelling, 152–153
  Resorption, 151–152
  Toxins and Rho, 153–162
Bordetella, 42
  Atrophic rhinitis, 25
  DNT, 33
Breast cancer
  Phytoestrogen protection against, 205
Brucella, 95
Burkholderia pseudomallei
  TTSS,

C2 toxin, 68
C3 toxin, 3, 154, 157
  Action on Rho, 37–38

Campylobacter
  CDT, 64
Cancer
  Bacterially induced, 4, 199–219
  ErbB2 overexpression, 58–59
  Helicobacter induction of, 169
  Molecular mechanisms of, 200
Carcinogen
  Helicobacter as, 169
Carcinogenesis
  Stages in, 200
Carrion's disease, 81, 86
Cat scratch disease, 82
  Differential diagnosis, 105
cbfa-1, 149
  Down regulation by PMT,
    157
CDC25 phosphotase
  CNF effect on, 59
Cdc42, 18, 33, 36
  Introduction, 14
Cell adhesion molecules
  Bartonella and, 91
Cell cycle, 55–59
  Control by CDTs, 65
Cell scattering
  Helicobacter induction of,
    179
Cellular Microbiology, iii, 53
Chemokines
  Helicobacter activation of, 172
Chronic infection
  Cancer and, 169
  Linked to inflammation, 4
CIF, 68, 228
Circular dichroism (CD), 10,
    23
Citrobacter rodentium
  CIF, 68
Citron kinase, 37
Clinical studies
  Helicobacter, 170
Clostridium difficile
  Osteoclasts treated with toxin,
    161
  Toxins, 154
  Toxins and Rho, 38

Clostridium novyi
Toxin and Rho, 38
Toxins, 154
Clostridium sordellii
Osteoclasts treated with toxin, 161
Toxins, 154
Toxins and Rho, 38
c-Met receptor tyrosine kinase, 180–181
Collagen, 148
Colorectal cancer
Bacteria and, 211–213
c-src
Activation by RANKL, 150
Cycasin
Carcinogen, 204
Cyclic AMP
Helicobacter activation of, 176
Cyclin-Dependent Kinases, 55
Cyclins, 55, 58
PMT and, 20
Rac, 36
Cyclooxygenase-2 (COX-2)
CNF stimulates, 42, 207
Helicobacter activation, 185
Helicobacter induction of, 175
MAPK activation of, 176
Cyclostatins, 53–74
Definition, 73
Experimental approaches, 70–73
Cytokines
Bone resorption, 147, 151
Cytolethal Distending Toxin, 53–74
Cytolethal Distending Toxin (CDT)
CDT-B, 60–62
CDT-B target, 61–62
CDT-C, 62
Colonisation by bacteria expressing, 64
DNA damage, 59–60
Effect on CDC25 phosphotase, 59
In vivo relevance, 63–65
Introduction, 53, 58–59

Cytotoxic Necrotizing Factor (CNF), 68, 154, 229
Apoptosis, 45
Cancer and, 45
Catalysis, 41, 44
COX-2 activation of, 42
Disease linked to, 44–45
E. coli urinary tract infection, 207
Introduction, 33
Lethality of, 39
Mode of action of, 40–41
Multinucleation caused by, 41
PMT homology, 22
Structure of, 39–40, 41, 43–44
Uptake, 39

Deamidation
Of Rho by CNF, 40–41
Of Rho by DNT, 42
Dermonecrotic toxin (DNT),
Bone loss due to, 155
Catalysis, 44
Disease linked to, 45, 154, 228
Effects on bone, 45
Introduction, 33
Multinucleation caused by, 42
Rho activation, 42–43
Structure, 43–44
Diaphanous, 34, 37
Differentiation
PMT inhibition of, 25
DNA
CDT target, 61–62

E. coli
Cif, 228
EspF and cell death, 135
Type III effector, 128
Endotoxin
Bone resorption and, 147
EPEC
Tir effector, 128
Epidermal Growth Factor (EGF)
Cell cycle and, 58

Epidermal Growth Factor (EGF) receptors
  Action by SAGP, 69
  Helicobacter activation of, 177–179
  PMT activation and, 19
  Transactivation independent of *Helicobacter*
    cag, 178
ErbB-2
  Helicobacter activation, 179
ERK1, 132
ERKs, 18
  And cell cycle, 58
  And PMT, 18–19
Escherichia coli
  Bladder infection, 207
  CDT, 59, 64
  CIF, 68
  CNF, 33
  STa, 69
  TTSS, 117
E-selectin, 91

F. nucleatum immunosuppressive protein
    (FIP), 69
Fas
  Helicobacter, 183
Fibronectin, 151
Focal adhesion kinase
  Activation by CNF, 41
  Activation by PMT, 15

Gallbladder
  Cancer, 213–215
  Cancer link with gallstone, 214
GAP activity
  ExoS, 127
Gastrin
  Elevation due to *Helicobacter*, 172
Gender differences
  Response to *Helicobacter*, 172
Glucosides
  Plant glucoside metabolism, 204
Glucosylation
  Of Rho proteins, 38
Glycosides
  Conjugates as carcinogens, 204–205
G-proteins
  Activation of Rho, 14

G_{12} family and PMT, 21
G_q, 156
G_q activation by PMT, 13–14, 18, 21
G_q activation of PLC-β, 12
  Introduction, 11–12
GTPase activating proteins (GAPs), 33
Guanine nucleotide exchange factors (GEFs),
    33
Guanylyl cyclase
  STa, 69

Haemolytic and uraemic syndrome (HUS)
  VT, 69
Haemophilus ducreyi
  CDT, 59, 64
Helicobacter, 169–186
  Animal models, 171, 174–175
  Animal models with *H. felis*, 171–172
  Animal models with *H. mustelae*, 171
  Apoptosis, 183–185
  cag, 172–173
  cag and c-Fos, 184
  CagA and cancer, 182
  CagA interaction with PLCγ, 181
  CagA target, 181
  Cancer, 4, 208–209
  CDT, 64
  Cell scattering, 179
  c-Met, 180–182
  COX-2 activation, 175, 185
  Effects on cell lines, 173
  EGF receptor activation, 177–179
  Eradication, 170
  Gastric cancer, 207
  Host factor involvement, 182–185
  Induction of proliferation *in vivo*,
    173–174
  Infection linked to blood group antigen
    babA2, 173–174
  MAPK activation, 182
  Mongolian gerbil animal model,
    174
  NF-κB, 174
  Oesophageal cancer, 215
  PIP, 70
  P13-K activation, 181
  Population effects, 208

Rho activation, 181
Signalling effects, 175–182
T cell involvement, 183
Virulence factors, 172–173
Hepatocyte growth factor (HGF), 180
Histidine decarboxylase
  Helicobacter activation of, 176
HUVECs, 17, 90
Hyperplasia
  Endothelial hyperplasia due to *Bartonella*,
  83
Hypochlorhydria, 209–211
  Bacterial growth due to, 207
  Pancreatic cancer, 218

ICAM-1, 91
  Bartonella effects on, 96
Inflammation
  Apoptosis and, 132
  Bartonella infection, 83
  Cancer and, 4, 185
  Helicobacter and, 170
Injectisome, 119
Inositol trisphosphate (IP$_3$)
  Calcium release, 12
Integrin
  Osteoclast adherence, 151
  Rho activation and signalling, 35

Jun kinase (JNK), 18, 132
  Activation by RANKL, 150
  CNF activation, 42
  Induction by *Salmonella* SopE, 122
  PMT and, 19
  Rac, 36
  VEGF activation, 101

Kaposi's sarcoma, 83
  Similarity of lesions to bacillary
  angiomatosis, 105

Leucine-rich repeat (LRR) proteins, 136
Leukaemia inhibitory factor (LIF), 70
Lignins
  Anticarcinogen, 205
Lung cancer
  Tuberculosis link, 218

Macrophage colony-stimulating factor
  (MCSF), 149
Matrix metalloproteinase-3 (MMP-3)
  Helicobacter increases, 178
Metabolites
  Carcinogenic, 202
Metalloprotease
  EGF receptor transactivation, 177
Mitogen-activated protein kinase (MAPK)
  Inhibition by SAGP, 69
Motogenic response
  Helicobacter, 179–182
  Helicobacter TFSS required for, 180
Mutation
  Role in cancer, 200
Mycotoxins, 201
  As anti-cancer agents, 201
Myocardial hypertrophy
  PMT and, 20

NF-κB, 132
  Activation by RANKL, 150
  Bartonella and, 98
  Helicobacter activation, 172, 174
  Introduction, 132
  Salmonella SspH1, 136
  Shigella IpaH, 98
N-nitroso compounds (NNC), 202–203
Normal bacterial flora
  Protective role, 199

Oesophageal cancer
  Helicobacter pylori link, 215
Oroya fever, 86
Osteoblasts, 25
  Introduction, 148–149
  PMT interaction with, 156–158
Osteocalcin, 149
  Down regulation by PMT, 157
Osteoclasts, 149–151
  PMT activation, 161
  PMT inhibition, 161
  PMT interaction with, 158–161
Osteopetrosis, 148
Osteopontin, 151
Osteoporosis, 148
Osteoprotegerin, 151

p53, 58
 Helicobacter and, 183, 184
PAK
 Activation by Bartonella, 98
Pancreatic cancer, 218–219
Pasteurella multicoda toxin (PMT)
 DNT homology, CNF homology, 43
Pasteurella multocida, 7, 25
Pasteurella multocida toxin (PMT), 7–27
 As a tool, 13, 20
 Atrophic rhinitis, 7, 24–25
 Bombesin interaction with, 12
 Bone loss due to, 155–161
 Cancer, 25
 Cell signalling, 10
 CNF homology, 22
 Domain architecture, 22–24
 Facilitation of bombesin signalling, 13
 G_{11} uninvolved, 14
 Gene, 22
 G_q activation, 13–14
 $G_{\alpha q}$ tyrosine phosphorylation induced by PMT, 21
 Intracellular action, 9–10
 Lethality, 25
 Mitogen, 7–9, 25
 Osteoblasts and, 156–158
 Osteoclasts and, 156–161
 Phospholipase C activation, 11–14
 Protein kinase C activation, 12, 20
 RANKL, 159
 Receptor, 10
 Rho activation, 16–17
 Rho and, 156
 Signalling and, 156
 Target, 21
 Transendothelial permeability, 17
 Variable mitogenicity, 20
Pathogenicity island (PAI)
 cag, 169
Peptic ulcer
 Cancer linked to, 209–210
Periodontal disease, 147
 A. actinomycetemcomitans, 64
Pernicious anaemia (PA)
 Cancer link, 209, 210

Peroxisome proliferator-activated receptor (PPAR), 174
 Helicobacter induced apoptosis, 184
Phospholipase
 Absence in C57Bl/6 mice, 171
 ExoU, 125
 P. aeruginosa, 135
 Phospholipase A_2 activation by Helicobacter, 176
Phytoestrogens, 205–206
PI3 kinases
 Cell cycle control, 58
 Rho activation, 35
Plants
 Anticarcinogens, 205–206
PMT
 See Pasteurella multocida toxin
Polycyclic aromatic hydrocarbon
 Detoxification, 204
Polycyclic aromatic hydrocarbon (PAH)
 As carcinogens, 204
Population
 Link to cancer incidence, 206
Prevotella intermedia, 70
Proliferation
 Helicobacter induced, 170
 Helicobacter induced proliferation and cag status, 173
 Rho, 35–37
Prostate
 Disease and CNF, 44–45
Prostate cancer
 Phytoestrogen protection against, 205
Prostatitis
 E. coli infection and, 207
Protein kinase C (PKC)
 Activation, 12
 Mycobacterial stimulation, 218
 PMT activation, 12
 PMT Erk phosphorylation, 19
Pseudomonas aeruginosa
 ExoS effector, 125–127
 ExoT effector, 127
 ExoU and cell death, 135
 ExoY, 136
 TTSS, 117

Ra1A
 Action of ExoS, 127
Rab
 Action of ExoS, 127
 Osteoclast function, 162
Rac, 18, 33
 Degradation, 42
 Introduction, 14
 PAK, 35
 Proliferation and, 36
 Rho antagonism, 35
Rap1B
 Action of ExoS, 127
Ras
 Action of ExoS, 127
 C. sordellii activation of, 38
 Introduction, 18
 Osteoclast survival, 162
Receptor activator of NF-κB (RANKL)
 PMT and, 159
Receptor activator of NF-κB ligand (RANKL),
  149
RGD, 151
Rho
 Bartonella and, 81, 92–95, 98–99
 Bone and, 153–162
 C3 target, 3
 Cell cycle, cell transformation, 34
 Cell transformation, 36
 GAP activity of YopE on RhoA, 124
 ICAM-1, 107
 Immunity and, 107–108
 Importance in bone resorption,
  161
 Introduction, 14–15, 33–34, 119–120,
  153–154
 Motility, role in 6
 Osteoblast differentiation, 158
 Osteoclasts and, 161–162
 PMT and, 16–17, 156
 Proliferation and, 35–37
 Regulation and adhesion, 107
 Rho A in erythrocytes, 97
 Rho effectors, 34–35
 Shigella type III effectors, 123–128
 Toxin targets, 227
 Toxins that activate, 39

 Toxins that inactivate, 37–38
 Type III effectors, 120
 VEGF activation, 101
Rho kinase, 15, 34–35
 Myosin light chain, 35
 PMT and, 16, 17
 Rac, 35
RNA interference (RNAi), 180

SAGP, 69
Salmonella, 120
 Effectors, 39
 SipA effector, 122
 SipB and apoptosis, 134–135
 SipB effector and cell death, 131
 SipC effector, 122
 SopB effector, 122
 SopB effector and inositol phosphatase,
  130–131
 SopE, 93
 SopE effector, 121
 SPI effectors, 121–122
 SPI-2 effectors, 136
 SptP action of JNK and ERK, 130
 SptP effector, 122, 130
 SspHs leucine rich repeat (LRR) proteins,
  136
 TTSS, 117
 TTSS effectors and signalling, 120–122
 TTSS effectors that affect Rho, 120–122
Salmonella typhi
 Carriers and cancer 9
 Infection and gallbladder cancer,
  214
Samonella
 Other type III effectors, 136
 SPI-2, 135
Severe combined immune deficiency disorcer
  (SCID), 183
Shigella, 120
 CDT, 59, 64
 IpaA, 123
 IpaB, 123, 131, 134
 IpaC, 123
 IpaHs leucine rich repeat (LRR) proteins,
  136
 IpgD, 131

Shigella (cont)
  TTSS, 117
  TTSS effectors that affect Rho, 123
  VirA effector, 123
Sickle-cell anemia, 91
Signalling
  Bone resorption and, 151–152
  Helicobacter induced, 169, 175–182
  Toxins targets, 3–4, 227
  Type III effectors that interfere with,
    119–136
Small interfering RNA (siRNA),
  180
Src
  Activation by PMT, 15
STa, 2, 69
STI, 70
Streptococcus pyogenes, 69
Structure
  CNF, 41

T cells
  Helicobacter induced pathology and, 183
Tartrate resistant acid phosphatase (TRAP),
  151
TNF-receptor-associated factor (TRAF)
  Activation by RANKL, 150
Toxin chimeras
  C3, 38
Toxins
  Bone and, 153–162
  Introduction and history, 1
Transendothelial permeability
  PMT and, 17
Transgenic mice
  Helicobacter research, 182
Transgenic mouse model
  Helicobacter research, 172
Transglutamination
  Rho transglutamination by DNT,
    43
Translocation
  Translocation via low pH step, 10
Transverse urea gradient gels, 10
Triple membrane passing signal (TMPS)
  EGF receptor activation, 177
  Over-expression in gastric cancer, 179

Tryptophan
  Metabolism linked to cancer, 202
TTSS
  Cell death, 131
Tuberculosis (TB)
  Lung cancer and, 218
Tumour promoters
  Toxins as, 229
Type III effectors
  Action on bacterial entry, 120–128
  Action on Rho, 119–128
  Of unknown function, 136
Type III secretion, 2–3, 117–136
  Introduction, 117–119
  kinases, 128–131
  phosphatases, 128–129
Type IV secretion
  Bartonella and, 95
  Helicobacter, 179
Type IV secretory
  Helicobacter, 173
Typhoid
  Cancer link, 215
  Carriers and pancreatic cancer, 219
Tyrosine phosphatases
  YopH, 129
Tyrosine phosphorylation
  Inhibitor as anticarcinogen, 205
  PMT induction, 15–16
  Tyrosine phosphorylation of $G_{\alpha q}$,
    21

Ulceration
  Helicobacter, 169
Urinary tract
  Cancer, 206–207
  Disease and CNF, 44–45

Vascular Epithelial Growth Factor (VEGF),
  100–106
  Apoptosis, 102
  Block of NF-κB, 105
  Endothelial mitogen, 101
  FAK Src, 102
  Hypoxia, 100
  Immune function, 104
  Inflammation, 104

Nitric oxide (NO), 103
p38 MAPK, 102
Paxillin, 102
Verotoxin (VT), 69
Verruga, 86
  Bartonella and, 82
Vitronectin, 151

Yersinia
  TTSS, 117
  Yops, 118–119
  YopE, 124
  YopH, 129–130

YopH and immune function,
  130
YopH interaction with P13K, 129
YopJ
  Apoptosis caused by, 132
YopJ and inflammation, 132–134
YopJ and signalling
YopJ/P and cell death, 131
YopM leucine-rich repeat (LRR) proteins,
  136
YopT effector, 124–125
YpkA/YopA effector, 131
Yops, 118–119